Animal Lives and Why They Matter

This book engages with the changing ways in which we, as a society and culture, look upon and interact with animals, stressing how much animals differ among themselves. An invitation to appreciate the peculiar role of animals in telling important if uncomfortable truths about who we are and where we are heading – namely, towards a world so much poorer in cultural, moral, and biological diversity – as a result of the ongoing decimation of so many other species. Drawing on a variety of thought ranging from that of Midgley, Plumwood, and Murdoch to Levinas, Derrida, and Habermas, from ecophilosophers to conservation biologists, *Animal Lives and Why They Matter* asks how we have come to this, and what an alternative, less destructive approach to our now precarious coexistence with animals might look like. Spanning the disciplines of philosophy, psychology, and anthropology, this enquiry into various cross-species relationships and encounters will appeal to scholars and students across the humanities and social sciences with interests in philosophy, ethics, human-animal interaction, and environmental thought.

Arne Johan Vetlesen is Professor of Philosophy at the University of Oslo, Norway. He is the author of *A Philosophy of Pain*; *Evil and Human Agency: Understanding Collective Evildoing*; *Perception, Empathy, and Judgment*; *The Denial of Nature: Environmental Philosophy in the Era of Global Capitalism*, and *Cosmologies of the Anthropocene: Panpsychism, Animism, and the Limits of Posthumanism.*

Multispecies Encounters

Series editors:

Samantha Hurn is Associate Professor in Anthropology, Director of the Exeter Anthrozoology as Symbiotic Ethics (EASE) working group, and Programme Director for the MA and PhD programmes in Anthrozoology at the University of Exeter, UK.

Chris Wilbert is Senior Lecturer in Tourism and Geography at the Lord Ashcroft International Business School at Anglia Ruskin University, UK.

Multispecies Encounters provides an interdisciplinary forum for the discussion, development and dissemination of research focused on encounters between members of different species. Re-evaluating our human relationships with other-than-human beings through an interrogation of the 'myth of human exceptionalism' which has structured (and limited) social thought for so long, the series presents work including multi-species ethnography, animal geographies and more-than-human approaches to research, in order not only better to understand the human condition, but also to situate us holistically, as human animals, within the global ecosystems we share with countless other living beings.

As such, the series expresses a commitment to the importance of giving balanced consideration to the experiences of all social actors involved in any given social interaction, with work advancing our theoretical knowledge and understanding of multi-species encounters and, where possible, exploring analytical frameworks which include ways or kinds of 'being' other than the human.

Titles in the series:

Blogging Wildlife: The Perception of Animals by Hikers on the Appalachian Trail
Kate Marx

Totemism and Human-Animal Relations in West Africa
Sharon Merz

Animal Lives and Why They Matter
Arne Johan Vetlesen

Cat People: Human-Cat Interrelatedness in the Cat Fancy
Emily Stone

The full list of titles for this series can be found here: https://www.routledge.com/Multispecies-Encounters/book-series/ASHSER1436

Animal Lives and Why They Matter

Arne Johan Vetlesen

Routledge
Taylor & Francis Group
LONDON AND NEW YORK

First published 2023
by Routledge
4 Park Square, Milton Park, Abingdon, Oxon OX14 4RN

and by Routledge
605 Third Avenue, New York, NY 10158

Routledge is an imprint of the Taylor & Francis Group, an informa business

British Library Cataloguing-in-Publication Data
A catalogue record for this book is available from the British Library

Library of Congress Cataloging-in-Publication Data
Names: Vetlesen, Arne Johan, 1960- author.
Title: Animal lives and why they matter / Arne Johan Vetlesen.
Description: Abingdon, Oxon ; New York, NY : Routledge, 2023. |
Series: Multispecies encounters | Includes bibliographical references.
Identifiers: LCCN 2022018133 (print) | LCCN 2022018134 (ebook) |
ISBN 9781032330013 (hardcover) | ISBN 9781032330020 (paperback) |
ISBN 9781003317692 (ebook)
Subjects: LCSH: Human-animal relationships--Moral and ethical aspects. | Animal welfare--Moral and ethical aspects.
Classification: LCC QL85 .V45 2023 (print) | LCC QL85 (ebook) |
DDC 591.5--dc23/eng/20220809
LC record available at https://lccn.loc.gov/2022018133
LC ebook record available at https://lccn.loc.gov/2022018134

ISBN: 978-1-032-33001-3 (hbk)
ISBN: 978-1-032-33002-0 (pbk)
ISBN: 978-1-003-31769-2 (ebk)

DOI: 10.4324/9781003317692

Typeset in Times New Roman
by KnowledgeWorks Global Ltd.

Contents

Preface

In my book *The Denial of Nature*, I set out to expand the sort of philosophical agenda my thinking had been committed – or should I say confined – to since my years as a student of Jürgen Habermas in Frankfurt in the late 1980s. My growing sense of impatience with the "human beings only" framework for doing philosophy in general, and ethics in particular, clearly reflected not only my personal misgivings about the largely self-imposed limitations of my profession. It also, and much more importantly, reflects the urgency of topics more-than-human in the wider world, topics such as global warming and the extinction of species, both largely human-caused.

In the sequel to *Denial*, entitled *Cosmologies of the Anthropocene*, I sought to further expand my initial expansion by incorporating studies in social anthropology (my first love in academia) so as to address what I have come to view as two intimately interconnected losses, that of biological diversity and that of cultural diversity. Too often treated separately, due to the unfortunate inertia of the traditional intellectual division of labour within the University between the social sciences on the one hand and the natural sciences on the other, I endeavoured to bring the analysis of the two – twin – losses together. Doing so landed me in the company of advocates of animism and panpsychism, unusual bedfellows for a contemporary philosopher.

The present book is the completion of what has turned out to be a trilogy. As one of the students in my course on environmental philosophy put it to me a couple of years ago: "It's time that you write about animals, isn't it?"

I want to be perfectly clear that I come to the subject of animals, and why they matter, in these times of species extinction and habitat loss, as the philosopher I will always be despite my forays into other disciplines and lines of inquiry. I am not an animal studies scholar. Nor do I aspire to reinvent myself as an expert in animal ethics. Again, I shall approach the work of fellow philosophers – such as Habermas, but also Emmanuel Levinas and Jacques Derrida – in a highly critical spirit. Yet at the same time I hope that my background as "not one of them" will prove fruitful to my discussion of the work of authors writing about animals from truly diverse vantage points, ranging from ethnography to ethology, from cultural history to psychoanalytic theory. Animals are complex, as are our stances to and relationships with them, thus demanding a truly multidisciplinary approach.

I am indebted to the persons who took the time to comment on my manuscript, either in whole or in part: Per Bjørn Foros, Trond Arnesen, Anne Sverdrup-Thygeson, Espen Hammer, and Jan-Olav Henriksen.

I am also grateful to the colleagues, from various disciplines, with whom I have discussed many of the issues treated here, as well as to the students who have constantly challenged me to raise the hard questions about animals and humans that I try to illuminate, rather than answer in a definitive manner: Knut Ivar Bjørlykhaug, Odin Lysaker, Sigurd Hverven, Zemir Popovac, Freya Mathews, Per Nortvedt, Erik Stänicke, Erling E. Guldbrandsen, Dag Schmedling Dramer, Maria Guzman-Gallegos, Martin Lee Mueller, Julianne Grovehagen, Martina Mercellova, and Sindre Brennhagen.

I dedicate this book to my friend, kindred spirit, sorely missed: Keith Tester.

Horten, Norway

Introduction

"There is nothing now but kestrel"

In his *Lectures on Ethics*, Immanuel Kant tells us that "so far as animals are concerned, we have no direct duties. Animals are not self-conscious and are there merely as a means to an end. That end is man. We can ask, 'Why do animals exist?' But to ask, 'Why does man exist?' is a meaningless question" (1930: 239).

Kant is in no doubt that animals are utterly different from humans, different in the morally crucial sense of classifying them as means and humans as ends. Their usefulness to humans provides the answer to the question why animals exist. Without humans, then, there would be no point to the existence of animals on earth. In holding this opinion, Kant takes is for granted that everything of importance and value about the existence of animals is inseparable from their coexistence with humans, inseparable in the strong sense of being derived from it. Outside the constellation with their "others" – us humans – their existence holds no philosophical interest for Kant. The fact that a considerable number of species of animals existed prior to the emergence of *homo sapiens* on planet Earth is not something that Kant dwells upon, placing them in a void where they have no way of mattering.

This is not to suggest that animals, precisely in their coexistence with humans taken for granted by Kant, are not important to him. In fact, their absence would make for a sort of moral deprivation. Why? Following the statement cited above, Kant observes that "We can judge the heart of a man by his treatment of animals"; indeed, humans should endeavor to be kind to animals, "for he who is cruel to animals becomes hard also in his dealings with men" (1930: 240). In other words, the presence of animals in the lives of humans is important in providing individual human persons with a relationship with a species "other", a relationship that is conducive to the cultivation of virtues appropriate in the dealings with fellow human persons, considered as the only carriers of moral status, of intrinsic worth. Animals are a means to this end, a go-between for the individuals who are the only genuine parties to a moral relationship.

In her seminal essay "The Sovereignty of Good over Other Concepts", the British novelist and philosopher Iris Murdoch, writing almost 200 years after Kant's lectures, shares the following experience with her readers:

DOI: 10.4324/9781003317692-1

> I am looking out of my window in an anxious and resentful state of mind, oblivious to my surroundings, brooding perhaps on some damage done to my prestige. Then suddenly I observe a hovering kestrel. In a moment everything is altered, The brooding self with its hurt vanity has disappeared. And when I return to thinking of the other matter it seems less important. [...] We take a self-forgetful pleasure in the sheer alien pointless independent existence of animals, birds, stones and trees.
> (Murdoch 1970: 84, 85)

Murdoch takes this experience as her starting point for exploring "moral change", holding it to be "the most accessible one", presumably because the experience is of a kind that will be eminently familiar to her readers. In speaking about moral change, Murdoch may seem to approach the importance of animals in the lives of humans in a way similar to Kant's, highlighting the cultivating role that animals may play for us in our capacity as moral agents. But this similarity is misleading. Before sharing her experience of seeing the kestrel, Murdoch submits that "beauty is the convenient and traditional name of something which art and nature share, and which gives a fairly clear sense to the idea of experience and change of consciousness" (1970: 84). Hers then is not first and foremost a moral interest in the kind of change that experiencing, say, an animal like a kestrel may bring about, but an *aesthetic* one, allowing an access to, and a deep appreciation of, beauty understood as "something which art and nature share".

To be sure, Kant is no foreigner to the affinity between art and nature that Murdoch points to, devoting an important section of his *Critique of Judgment* to it under the heading "the sublime". "Nature is beautiful", writes Kant, "because it looks like art, and art can only be called beautiful if we are conscious of it as art while yet it looks like nature" (1951 [1790]:149). But individual animals like Murdoch's kestrel play no vital part in Kant's account. And in fact, Murdoch too soon moves to other examples than those provided by animals. "Good art", she contends, "reveals what we are usually too selfish and too timid to recognize, the minute and absolute random detail of the world, and reveals it together with a sense of unity and form"; it "shows us how difficult it is to be objective by showing us how differently the world looks to an objective vision". Good art, then, "is something pre-eminently outside us and resistant to our consciousness. We surrender ourselves to its *authority* with a love which is unpossessive and unselfish" (1970: 86, 88). Being what Murdoch describes as "a kind of goodness by proxy", art helps us "transcend selfish and obsessive limitations of personality", thus enlarging the sensibility of the human subject (87). *Attention*, or looking, looking properly and deeply, is absolutely crucial for Murdoch, not only to grasp beauty but to orient oneself morally, to grasp and do justice to what really matters: "If I attend properly I will have no choices and this is the ultimate condition to be aimed at" (40). In this spirit, inspired by Plato, virtue is "*au fond* the same in the artist as in the good man in that it is a selfless attention to nature", showing art and morals to be "two aspects of a single struggle" rather than in conflict (41).

What became of the "self-forgetful pleasure" that we according to Murdoch take in "the sheer alien pointless independent existence of animals, birds, stones and tress"? What happened to the animals?

Before trying to answer that question, notice a big difference between what Murdoch and Kant say about animals: Murdoch does not hold their not being self-conscious against them like Kant does, using that alleged deficit as a justification for viewing animals as merely a means to an end, that is, the end(s) represented and pursued by humans and humans only. Quite the contrary, in fact: the kestrel that Murdoch sees hovering when looking out of her window exemplifies "the sheer alien pointless independent existence" of animals in general. That the kestrel's existence is aptly described in these exact terms is what we are invited to think of as what endows it with the value Murdoch recognizes in it, as inseparable from it, so that if we don't see these features as meriting our admiration, our self-forgetful pleasure in their sheer existence, Murdoch will urge us to look again, and again, if necessary.

The deeper meaning of the difference between their views is more subtle, though. It is not that Murdoch refuses to go along with Kant in holding animals' lack of being self-conscious against them; rather, it is that the self-consciousness or lack of such of animals does not enter Murdoch's experience, and her valuation of that experience, at all. What makes such a big and morally crucial difference to Kant makes no difference whatsoever to Murdoch. Does she think that the kestrel she sees is self-conscious? We don't know – she doesn't say. Her not saying means that the possession or lack of such of this particular property is not part of what is important about the kestrel. What is important, then? Its sheer alien pointless independent existence.

By far, I take the most intriguing part of Murdoch's description to be her matter-of-factly remark: "There is nothing now but kestrel." No further explanation is offered; it is as if this "nothing but" is self-explanatory. Having turned her attention, meaning full-fledged, concentrated attention, to the kestrel, it commands all Murdoch's powers of attention: the possession of *that ability* is the gift it is endowed with, being the crux of what an animal like a kestrel can do, can be, once given the chance. And since it has, in and of itself, what it takes to command attention in this profound fashion, we may speak about a before and an after: there is Murdoch's state of mind before noticing the bird outside her window, and there is her state of mind afterwards, being an altogether altered one.

Mary Midgley, philosopher and lifelong friend of Murdoch's, and a key figure in this book, makes a comment worth mentioning here: "If we found that we were in Disneyland, with plastic kestrels going up at carefully randomized intervals, [Murdoch's] entire point would be lost" (1995: 346). Evidently, Midgley sees no need to explain, let alone argue her point: it goes without saying that the distinction between the kestrel observed by

Murdoch and artificial ones is clear-cut and compelling for everyone and that real kestrels such as encountered in nature possess an ability to instil awe and admiration in us that is wholly lacking in kestrels made of plastic.

Philosophically, the distinction in play derives from the venerable one known to us since Plato and Aristotle, that between the grown and the made, whereby the grown – say, a bird such as a kestrel – exhibits an independence of being vis-à-vis us humans not found in the made, having been brought into existence by us and in order to serve some human-oriented purpose, as exemplified in the making of plastic kestrels: their very existence depends on us. True, we may admire what we have made, whereby works of art (Murdoch) certainly strike us as more worthy objects than, say, plastic kestrels. In either case, however, the admiration for such objects entails in no small measure admiration for their makers: look and appreciate what humans are capable of creating.

This is the dimension I take Murdoch to throw into sharp relief when picking the exact words cited above: "the sheer alien pointless independent existence of animals, birds, stones and trees". The selected words all underscore that the things she directs our attention to here are precisely not of our making – this holds not only for their sheer existence, but – fundamentally – for *the way* in which they exist, the *how* of their distinct being in the world. True, she adds a thought I haven't cited yet, namely: "Not how the world is, but that it is, is the mystical" (85). That may well be, but for my purposes the *how* – say, of a bird like the kestrel – is not secondary in importance to the *that*.

As for the how of animals, one may question whether the words of description used by Murdoch are as "objective" as her version of Plato-inspired objectivism aspires to be. Can we really say that the existence of animals is "alien"? Or "pointless"? Alien and pointless for whom, from whose point of view? Surely depicting their existence like this presupposes a particular point of view, particular in a sense giving the lie to the ambition of objectivity – namely, the point of view peculiar to humans. As for the animals themselves – can we say that their existence is alien, is pointless, to them? Shall we reply that "it depends", depends on their possession of specific capacities, such as self-consciousness (Kant), in which case we may be willing to grant that *some* animals are able to experience their existence as meaningful (if that is the alternative to pointless), whereas others are not? And what about "alien"? If anything, isn't the existence of non-human animals like birds the most natural thing there is? If depicted as alien, doesn't that say something about us and our particular point of view, and nothing about the birds?

This may sound abstract and academic, but it's not. The conditions under which animals live their lives are rapidly and drastically changing. Philosophers' predilection for examining their selected entities in a once-and-for-all manner, *sub specie aeternitatis*, comes up short in a situation like the present one, affecting not only *how* animals live their lives, but increasingly also *whether* the form of life they represent will go on existing at all.

Notwithstanding the many quarrels among philosophers over the correct description of properties taken to distinguish animals from humans, or one type of animal from others, the sheer continued presence of animals on earth is being taken for granted, unanimously. The constellation humans/animals, empirically and not only analytically understood, has been a constant, allowing for the assessment of alterations among the factors that *do* change, such as the geographical and numerical distribution of animals classified as wild, domestic, and pets. *That there will be animals as long as there are humans*, providing us with direct experiential access to our so-called "others", has never been in doubt – doubting it would seem so fanciful as to not be worth the exercise.

These days, it is not our inability to doubt the possible future absence of animals that deserves attention, but our inability – unwillingness – to recognize their disappearance even when staring it in the face. Perhaps this inability is not first and foremost a philosophical phenomenon but rather a psychological, sociological, or ideological one, suitable to the sort of inquiries that the disciplines thus alluded to will help us undertake. If we have difficulties recognizing the novel historical fact that animals – some types of them more than others, and some more rapidly than others – are disappearing, denial might be a better term than inability, as borne out in the title of my previous book *The Denial of Nature* (2015). Or perhaps we need to take a more long-term view on how it has come about that non-human "others" such as animals are seen as acceptable casualties of what we, generally speaking, welcome as progress, making the losses occurring along the way worth it, if not for the affected animals, then for us humans – a notion at the heart of the book preceding this one, *Cosmologies of the Anthropocene* (2019).

What was Kant thinking when he said that animals "are there merely as a means to an end" and "that end is man"? Are we now witnessing the utterly concrete consequences of acting – collectively, as a culture and society – in accordance with his matter-of-factly sounding assertion, one that drew little, if any, critical attention at the time? Or would putting it like this, in terms of a charge involving some portion of responsibility, be to do Kant an injustice, attributing an importance to his philosophical view about the relationship between humans and animals that is simply inappropriate, greatly exaggerating its historical impact? More to the point is perhaps to observe that he did not, and could not, foresee the situation we're in, having for centuries treated most animals as are they merely a means to our human ends, dispensable at that. While kindness to animals reflects the existence of a properly social and moral individual, according to Kant, viewing and treating them in that spirit and for that purpose leaves them without any proper protection, philosophically as well as practically.

In speaking about the disappearance of animals, showing beyond reasonable doubt that their future existence can no longer be taken for granted, what precisely am I referring to?

No longer is there anything new under the sun about newspaper articles that inform us that "researchers warned last year that two-thirds of North American bird species are at risk of extinction", explaining that "climate change is disrupting hundreds of bird species" (*The Guardian*, Weekly, January 15, 2021: 31). A comprehensive study led by Professor of Ecology Gerardo Ceballos, using a sample of 27,600 terrestrial vertebrate species, and a more detailed analysis of 177 mammal species, published in *Proceedings of the National Academy of Sciences* in July 2017, offers the following summary: "The biological annihilation we now observe will without doubt have severe ecological, economic, and social consequences. Humanity will eventually pay an exceptionally high price for the ongoing decimation of the only habitat we know in the universe. The situation is so critical that it would be unethical not to use strong words" (Ceballos et al. 2017: 3). The scientists continue:

> Dwindling population sizes and range shrinkages amount to a massive anthropogenic erosion of biodiversity and of the ecosystem services essential to civilization. [...] The ultimate drivers of the immediate causes of biotic destruction [are] human overpopulation and overconsumption, especially by the rich. These drivers, all of which trace to the fiction that perpetual growth can occur on a finite planet, are themselves increasing rapidly. All signs point to ever more powerful assaults on biodiversity in the next two decades, painting a dismal picture of the future of life, including human life.
>
> (Ceballos et al. 2017: 7, 8)

Thus the disappearance of animals that I spoke about above is part of a bigger picture: *biological annihilation*, a novel term marking a novel historical situation. With regard to animals, this means their ongoing decimation, eventually extinction. That's where the scientists tell us we are heading.

Since this is a book written by a philosopher, not a natural scientist, I shall not concentrate on numbers, knowing full well that some sorts of animals see a numerical increase (say, livestock), while others are becoming so few and far between as to be on the brink of extinction – partly because losing their habitat to the ever-expanding areas cleared for livestock. Indeed, to talk in general fashion about "animals" is hopelessly inaccurate. Not only do various sorts of animals exhibit an enormous heterogeneity as far as their capacities and ways of moving and being in the world are concerned. Different animals also face very different futures, often as a result of how differently some of them are treated by humans as compared with other animals. The arbitrariness to which we subject animals as members of what we hail as a scientifically informed, rationality-oriented society is staggering – if anything, it is more conspicuous today than in earlier epochs and in non-Western cultures, not less. Melanie Challenger puts it well: "The golden retriever that sleeps at the foot of someone's bed is no more intelligent or special than the pig they may have eaten earlier that day. But the retriever is in partnership with a human. What the dog and the pig have in common is that their worth has been arbitrarily given on the basis of the relationship a human wants with them" (2021: 90). Literally, the way the human party defines the relationship is nothing less than a life and death issue for the animal in question: "The dog and the man are friends, and so the dog is given identity and a mind, of sorts. The pig, just as sentient and sensitive a creature, can be rendered unconscious, hung from an overhead rail and slit behind their jowl to sever both jugular and carotid" (ibid.).

But it's not only that we are notoriously inconsistent in the way we discriminate between animals, maintaining best-friends attachment to the comparatively less intelligent retriever and having the more intelligent pig for dinner, making a mockery of the long-standing notion that intelligence is what counts. At a more basic level, the division that matters most and that never ceases to cut the deepest, both ways, is the generic and general one: humans versus animals, us versus them. As Challenger observes in *How to Be Animal*, "many people don't like to be reminded that we are animals, so much so that we may unconsciously rebel against anything that seems to place us in this category" (2021: 177). At bottom, then, the differentiation that really matters is *not in degree but in kind*, thus reversing Darwin's famous formula, as we shall see: it is an either/or, being human or being animal, as were the two mutually exclusive. If progress, the onward march of civilization, is about overcoming the limits of nature, attaining mastery over natural forces we previously were at the mercy of, then we will become increasingly intolerant in the face of conditions to do with the aging, ultimately the death of our bodies, intolerant when faced with the frailty of our existence as vulnerable and mortal beings, that is, as beings that share these features with our species others, animals, meaning with what we are so keen to maintain we are not.

But, we may ask, what is specifically human, if not the ambition to escape the conditions reserved for all *other* Earthborn creatures, animals in particular? Animals are stuck with their natural being, are they not? Stuck with and forever condemned to their ever-renewed but essentially static live and let die, to their mortality in particular, to given conditions in a way precluding any attempt to break free from them. Nourished by an ever-increasing sense of mastery, of expanding the range of control thanks to advances in science and technology, we modern humans see less reason than ever to surrender to natural forces, indeed to surrender anything at all. The result is that the perceived difference between us and animals – dynamic us, static them; culture us, nature them – grows bigger and deeper. In this perspective, the whole point about animals being radically different from us and everything that we consider most distinctive and most important about us – intelligence in particular – works normatively and not just descriptively: the differences made to matter most are always and invariably to the detriment of the animals.

Why does it matter whether we look upon animals first and foremost through the lens of difference from us or through the lens of similarity? Factually, both approaches will get some things right and overlook others. If differences appear to outweigh similarities, what follows? Traditionally, within Western thought and mentality, differences are to our gain qua humans, showing up what animals lack that we have, advanced cognitive abilities being the prime example. But why should the fact that birds (all sorts of them) are less adept than us humans at discussing abstract ideas and writing novels count as more important than the fact that birds can fly and we cannot? That they can navigate, down to the exact square meter, and travel between continents thousands of miles apart? If anything, isn't the arbitrariness shown in how we compare various types of animals with ourselves even more striking than the arbitrariness mentioned above in how we compare one sort of animal with another, such as the dog we call "man's best friend" and the pig we have for dinner? And isn't the first arbitrariness even more questionable than the second, in that it seeks to establish a categorical division between humans and (all sorts of) animals of utmost moral, legal, and political import, namely in reserving the status "end in in themselves" (or inherent worth) to humans, denying it to animals, notwithstanding how they excel in being capable of doing things we cannot?

Of course, in saying this I make a point that has been made over and over again since Jeremy Bentham's assertion that "The question is not Can they reason? Nor, Can they talk? But, Can they suffer?" (1948 [1789]: 311), inspiring Peter Singer's ground-breaking *Animal Liberation*. Even so, this book will pursue a different approach than the capacities-oriented one central to the discussion that Singer's modern classic has helped define for almost 50 years. Although I agree that *if* we attempt to assess the moral status of animals in terms of capacities, suffering is by far a better candidate than "reason" or speech, my interest in animals lies elsewhere.

To see where this leads, consider again Murdoch's statement about "the sheer alien pointless independent existence of animals, birds, stones and trees". As indicated above, to speak of the existence of animals as "alien" and "pointless" betrays a peculiarly human point of view, yet one less intuitively true about how humans view animals than Murdoch seems to think. Even if we grant that people in the contemporary Western society Murdoch takes to be her audience consider the existence of animals "alien" and "pointless", their doing so is far from a self-evident fact. Rather, it makes for a conspicuous break with how animals have been viewed in Western societies in the past, besides making for an even more striking contrast with how animals have been, and still are, perceived in non-Western cultures studied by anthropologists and for the most part ignored by philosophers.

Murdoch shows no interest in how animals view their own existence. How plausible is it that they will agree with Murdoch's human, all-too-human depiction of it as something "alien" and "pointless"? Even if it strikes humans (or some of us) as alien, can it really be held to be so to the animals themselves?

Hardly. If anything, the existence of humans, and the ways in which we conduct ourselves, lacking such an elementary ability as that of flying, will appear alien to the animals, especially perhaps, by way of comparison, to birds, the one type of animals mentioned specifically by Murdoch. Speaking from experience, I am in no doubt that, say, the crow in my garden finds my existence and ways of behaving alien, full of contradictions and raising lots of questions about the possession of the kinds of capacities we identify (in ourselves only) as reason and advanced intelligence; whereas for other sorts of animals, such as the retriever, our behaviour qua humans will appear as the most familiar thing in the world, inseparable from their lifeworld qua dogs.

This is not only a reminder of the massive limits to talking about "animals" in a general sense. To observe how enormously animals differ among themselves when it comes to what defines the world(s) they inhabit is important in its own right, showing the limits to the relative-to humans perspective that Murdoch exemplifies, understood as the rule not the exception. Having said this, Murdoch clearly *values* the existence of animals as something "alien" and "pointless", experiencing it as something genuinely positive – that their existence strikes us like this does not detract from its splendour, its enriching the world. To the contrary, being – from a human point of view – pointless may be said to be *the whole point* about there being animals in the world.

Why is it that the whole point of the existence of animals is borne out by their striking us as pointless in Murdoch's sense? What does pointless mean? One possible answer is: without purpose. Or to be more precise: without a purpose accessible to, graspable for, us humans, or (more pragmatically put) without a purpose that seems useful to *us*, thus manifesting *a limit on our part* and not on that of animals, *a limit to what we may hope to understand and to what we recognize as useful.* There is, I think, something admirable about this view in that it departs from the culturally entrenched notion that the possession of a purpose, or the role in helping realize some purpose (be it one that is more manifest in *other* entities or creatures), is a requirement for having value and for meriting respect for what it – say, a kestrel or a crow – is in its own right. Indeed, the point can be put more strongly still: that there be, and continue to exist, creatures in the world that are admirable and valuable in themselves irrespective of and independently from the means-ends framework to which the notion of purpose is wedded (at least in modern Western philosophy and cultural mentality) – *that* is what is truly important about them, meaning animals in this case.

I think that this is a better answer to Kant's question "Why do animals exist?" than the one given by Kant in his lectures on ethics, tied as it is to the means-end nexus, unable and unwilling to grant animals an importance other than being means to human ends. What is more, the question Kant considers meaningless – "Why does man exist?" – is a very good one, not only from the point of view of the animals treated as mere means by this relatively recent species on planet Earth, but also, I will argue, for a growing number of humans today. Indeed, the question why *we* exist, understood not only as a metaphysical one that in a sense must remain speculative, but also as a question about our *moral* status, a status very much borne out in how we act upon a denial to accord that status to any other creatures than ourselves, is presently assuming a meaning and an urgency Kant could not anticipate. Let me put it like this: the way we as a society have treated animals in the recent past compels us to pose anew the question of *who we are*, and of *what our role as one species among so many others should be*, having seen ourselves as entitled to a set of behaviours that have led to the biological annihilation described by Ceballos and his colleagues.

Having commented on Murdoch's use of the words "alien" and "pointless" to describe the existence of animals, something must be said about the third concept she uses, namely "independent".

For my purposes in this book, the issue of independence is more important than the other two, being less metaphysical and more literal. Now if I say that in my perspective the main thing about animals is their independence from us humans, the response will probably be that in taking such a view I greatly, and problematically, restrict the kinds of animals I am talking about, qualitatively as well as numerically. Well, that is exactly what I seek to do, in keeping with my critical remarks with regard to approaching animals as were they all the same as far as the question why they matter is concerned.

Of course, to point of how much different sorts of animals differ has become commonplace. And yet the three-partite categorization that seems so self-evident to us today, exemplified in the types of animals we may encounter within a single day – namely wild, domestic, and pets – is nothing of the sort, but itself a product of profound changes within Western society during the last couple of centuries, changes that provide the sources I rely upon throughout this book. For fear of putting the major historical trends in too simplistic a manner, it is essentially a story about how animals that used to be abundantly wild – in the abundant wild – came to be domesticated, and how (some) domestic animals in their turn have increasingly become pets. Indeed, in the last chapter we will see that one of the most striking signs of the trend I sketch here consists in how even the few animals left in the wild nowadays tend to be turned into pets, if not literally then in terms of how they are perceived, making them subject to a sentimentality that I will argue sits uneasily with their nature as wild animals.

Given this situation pertaining to so many animals today, what do I mean by their "independence"? If the point of their existence is inseparable from their independence – speaking of some animals, not all, of course – what about the ecological argument that all creatures play a role in the existence of (some, not all) others, exhibiting their inter-dependence, cross- and inter-species at that, and giving the lie to the notion that there is such a thing as a truly independent creature?

I do not deny the ecological argument thus conceived. In fact, it is part of my own argument, as brought out in my agreement with the holistic biocentrism worked out by environmental philosopher Holmes Rolston, demonstrating the continuity between my two previous books – *Denial* and *Cosmologies* – and this one. To get this, it is important not to conflate different perspectives and levels of analysis. In this book, I endorse the ecological argument about the essential inter-dependency and connectedness between different species, including different sorts of animals. No animal is an island. Hence I follow Rolston's holistic biocentrism in holding that, put formula-like, the morally right and the ecologically required are but two sides of the same coin, meaning mutually supportive – in the big picture and in the long run, that is. This does not preclude conflict between the two in the concrete case, that is to say, a particular place involving particular animals; in Chapter 5 we will look at several cases in point. Indeed, moral *and* ecological dilemmas unknown until recently today impose themselves as a direct result of the biological annihilation mentioned above, raising novel questions in a novel situation, questions about "sacrificial logic", about deciding that some animals must die in order that others be saved from extinction; about the validity of the once clear-cut but now blurred distinction between animals *in situ* and *ex situ*, in their original habitat and in around-the-clock human-controlled artificial environments (zoos, labs); and about the appropriateness of holding on to a notion of wild animals, and of wilderness, in times of human-made impacts on the conditions of animals and the ecologies they are part of virtually everywhere on planet Earth. Practices of hunting, as still found in a number of indigenous cultures, play a major part in my argument here, challenging the view dominant among animal rights theorists, Singerian or otherwise, that killing animals is wrong, period. For me, the manner of death is not the primary issue; the place allowed for the exercise of the full powers of agency of the animals in question is.

To be sure, then, no animal is an island. But some animals can do without *us*. They even have a right to do – to live their lives – without us. *This* captures what I mean by their independence, and I guess that Murdoch would agree, her kestrel being just as apt an example as the eagle, the wolf, and the saltwater crocodile. Speaking of their being independent creatures like this, independent from us, does not contradict their being ecologically interdependent and interconnected, with various other species of animals, as part of a wider ecosystem tied to a particular place, in the sense stressed in the biocentric outlook I endorse. As for us humans, the problem we may have with allowing these animals to retain and to flourish in their independence vis-à-vis our human ways and activities, expanding everywhere as they currently do, is in my view tied to our reluctance to recognize that *we* are the dependent party, not they. How so? Ultimately we depend on non-human species and forms of life – including animals – in order to survive. In saying this, I wish to leave aside the question of whether or not we may, if only we decide to, live without killing and eating animals. My point is a different one, inspired by Murdoch: the human dependence on animals that interests me the most is to do with quality of life; with how the quality of life of Murdoch's kestrel enriches not only that bird's life but Murdoch's as well, whereby the bird's independence from her, it being different, neither instructed nor controlled by her, is the precondition as well as the core of that quality.

I shall move back and forth between a macro- and a micro-perspective to illuminate what relationships between humans and animals have come to be like in present-day society, drawing on cultural history and anthropology to help us appreciate the substantial changes that have occurred – changes entailing, as their end point, the decimation and ultimately extinction of a record number of species of animals at record-breaking speed. The biological annihilation that Ceballos directs our attention to serves as a backdrop that conveys a sense of alarm and urgency. But rather than being my main topic, the ongoing anthropogenic degradation of the natural world that causes annihilation provides resonance for my overall emphasis on encounters between humans and a particular animal; encounters that, when discussed by philosophers, tend to be generalized in ways not justified by their being just that: encounters with *particular* animals. Encounter is a better word than relationship, because the latter carries assumptions about reciprocity, actual or possible such, in a way that an encounter does not. To the extent that the encounters I talk about may be placed on a continuum from human-animal dependence to independence – with, say, a retriever exemplifying the one end and a saltwater crocodile the other – I shall have more to say about the latter cases than the former, not least because of their rarity and precariousness.

This being my approach, the animals that figure prominently in the chapters that follow are the following: a group of baboons; a dog named Bobby; dingoes (Australian wild dogs); a cat looking at a naked and very famous philosopher; a crocodile perceiving its (his or her) human counterpart as food; a horse called Blue; and last but not least, Little Penguins trying to find their home. The fact that they constitute such a diverse sample is important and in keeping with my ambition to do justice to how richly different animals are, and the limits thus posed to philosophers' predilection for generalization.

In discussing Kant, Mary Midgley remarks that when philosophers announce that they will be talking about animals, they end up talking about humans. I am well aware that this book is no exception. I happen to be a philosopher, and I claim no expert knowledge in the field of animal studies, even though much of the material examined, and the authors whose views I discuss, belong to that field, especially – and tellingly – to the recently emerging one called extinction studies. One of the most fascinating things about writing this book has been to discover just how much a given philosopher's views on animals reveal about his or her views on the matters that matter most to humans – animals not really being admitted as part of those matters, but coming in handy to shed light on them nonetheless. For example, Descartes' statements about animals inform us just as accurately about his ethical position as his statements about humans, if not more; the same may be said, though perhaps more controversially, about Kant. There is a kind of consistency to this in that the non-human creatures who are considered to be mere means to human ends, are treated as mere means analytically no less than in real life: to make a point about humans, not animals; points to some extent parasitic on our declared "others", showing who we are by way of who we are not. Consistency is a virtue, even a trump virtue, in philosophy, but the price paid by those who are consistently treated as mere means in practice as well as in theory, is growing unacceptably high – not only for the animals affected.

Let me offer a brief overview of the separate chapters.

In Chapter 1 I discuss Mary Midgley's attempt to answer the question "why animals matter", leaning heavily on the work of ethologists such as Konrad Lorenz, Niko Tinbergen, and Jane Goodall. The notions of anthropomorphism and of animal mind are central to Midgley's approach, raising difficult epistemological questions about how to go about understanding the world in which animals live, being similar to us in some respects, radically different in others. How to treat the element of cross-species difference – as a regrettable limit to how much we may know about animals' perspective on the world, or as the major point about who they are? I then go on to examine the work of the American philosopher Paul Shepard, dedicated to demonstrating how it is that animals make us human. Such "making" is a process as crucial as it is precarious, embedded in changing cultural circumstances, sometimes conducive to how animals as our "Others" in Shepard's emphatic sense make us who we are, other times undermining the potential for human growth and maturity by robbing animals of the otherness that, again, is the major point, domesticizing them and turning them into pets. And make no mistake: whereas we do this to our detriment as humans, the wrong it entails is a wrong done to the animals.

In Chapter 2 I present philosopher Cynthia Willett's proposal for an "interspecies ethics". Using call and response as a model for interplay and communication across species, Willett's work highlights the rich potential she sees for a genuinely moral, and to some extent reciprocal, symmetric, relationship between humans and various sorts of animals. Although it is easy to be in sympathy with this approach, and although it contains important insights, I point out its blind spots and limitations. I bring some of these to our attention by looking at philosopher and anthropologist Deborah Bird Rose's reflections on the plight of dingoes in Australia. Central to Rose's discussion is a practice passed over in silence in Willett, namely hunting. This of course raises the question of the rightness, or wrongness, of humans' killing animals, of taking part, across species, in the reality of eating and being eaten.

Chapter 3 is devoted to the views on the moral standing of animals articulated by three giants of late twentieth-century philosophy: Jürgen Habermas, Jacques Derrida, and Emmanuel Levinas. In talking about animals (his cat in Derrida's case, the dog Bobby in Levinas'), all three of them tell us something vital about their overall understanding of what is required to qualify as a moral agent and as a moral addressee, and how the relationship formed by the parties to it should be considered: in terms of symmetry or of asymmetry? The question with which Levinas in particular is associated – "Does the animal have a face?" – will be central to the discussion.

In Chapter 4 I turn to Val Plumwood's encounter with a crocodile, an encounter she was lucky to survive, marking a before and after in her reflections about human-animal relationships. Rarely does it happen that the prospect of being eaten imposes itself with such existential depth on a philosopher taking part in the otherwise academic debate over predator and prey, where the one preyed upon is always an animal, never a human, let alone oneself. Plumwood's experience and her reflections on it set the stage for my discussion of the work of American moral philosopher Gary Steiner, a work that seeks to provide a principal argument for why killing – and eating – animals is wrong, period. I offer my objections to Steiner's argument before examining how author Alice Walker's relationship with a horse named Blue throws up questions of responsibility and loyalty in a situation of crisis for the horse.

Finally, in Chapter 5, I address the ways in which the danger of going extinct facing many species of animals today force us to reconsider the ethics as well as the appropriateness of established practices of conservation, be it *in situ* or *ex situ*, if we accept that this distinction still holds – which I do, as part of my case for upholding the notion of wilderness. I discuss the concerns of extinction studies, such as the dilemmas involved in the "logic of sacrifice": using some animals as means for the survival of others. And I end by asking the question Judith Butler raised years ago, about whose lives are considered grievable and whose not. Since the extinction I am concerned with is that of animals, this question leads to a reflection on ecogrief – or ecological mourning – as a response to the largely human-caused peril in which so many animals now find themselves.

1 Animals make us human

I

In the introduction to her first book, *Beast and Man*, published in 1978, Mary Midgley writes: "We are not just rather like animals; we *are* animals. Our difference from other species may be striking, but comparisons with them have always been, and must be, crucial to our view of ourselves" (1995 [1978]: xxxiii). Animals, Midgley goes on to point out, "are not machines". Indeed, one of her main concerns is "to combat this notion", stressing that "actually only machines are machines" (xxxvi). It follows that Descartes was wrong in holding that animals are "automata"; when he did so, his premise was the sort of absolute division between humans and animals that Midgley opposes, suggesting a continuum instead. "Man", Midgley writes (whereby she means, true to the tradition she wants to challenge where the masculine view has prevailed, human beings), "has his own nature, not that of any other species". Crucially, man is "*more* free than other species. But that extra freedom flows from something natural to him – his special kind of intelligence and the character traits that go with it" (xxxvii).

The notion that freedom flows from "something natural" in humans is one that many philosophers have found problematic. Midgley is acutely aware of this, and – shifting the focus from the historical past to the present – she mentions the dispute prompted by the publication of Edward O. Wilson's influential book *Sociobiology: The New Synthesis* in 1975, where some of the strongest opposition came from people who believe that "any notion of inborn active and social tendencies, if extended to man, threatens human freedom" (xxxvii). To Midgley, the idea that humans can be free only inasmuch as we are set radically apart from nature and exhibit traits unique to us among all species is plain wrong. Far from threatening the concept of freedom, then, to hold that we "have a nature" is absolutely essential to that freedom. "If we were genuinely plastic and indeterminate at birth", argues Midgley, "there could be no reason why society should not stamp us into any shape that might suit it" (ibid.).

The blatant denial that "man" has a nature is found in Existentialism, particularly in the works of early Sartre, predicated as Midgley sees them to

DOI: 10.4324/9781003317692-2

be on the notion that nature and freedom are mutually exclusive. "There is no human nature", she quotes Sartre famously proclaiming in *Existentialism and Humanism*, his most popular book, meant for a non-philosophical audience that proved more than willing to embrace its message in post-war France in the late 1940s. "Man first of all exists", Sartre contends, "he encounters himself, surges up in the world, and defines himself afterwards. If man as the Existentialist sees him is not definable, it is because to begin with he is nothing. He will not be anything until later, and then he will be what he makes himself" (Sartre 1958: 28). In other words, humans become human, and exhibit their unique capacity for freedom, only by way of being different from all other creatures. Nature as such is philosophically of no interest to Sartre; this holds for his major statement *Being and Nothingness* (1943) no less than for his non-academic writings, including plays and novels. Everything of importance about the human condition has to do with that condition being something emphatically cultural and social, not natural. To take recourse to something putatively "natural" in order to explain a piece of human behaviour would immediately – it is tempting to say: instinctively – raise suspicion about being in "bad faith", appealing to the workings of forces of determinacy such that one cannot be held responsible for what are truly choices freely made. In Sartre no less than in Kant before him, then, nature designates the sphere of heteronomy and thus the dimension human agents need to negate, or transcend, to realize their true humanity.

It goes without saying that there is no need for animals in this outlook; no need, that is, to acknowledge even the sheer presence on earth of animals in order to say what essentially needs to be said about how humans come to be human and epitomize their unique species in each individual case. Processes of socialization, the stages required to undertake the ontogenetic trajectory from infant over child and adolescent to the full-grown adult – this can and indeed should occur without any significant role being played by non-human creatures in the form of animals. Picking Sartre as the modern-day champion of this view, while justified, risks overstating the extent to which this view qualifies as distinctly *modern*. The truth is that the view goes all the way back to Socrates and hence to Plato. It was Socrates who declared that in order to experience anything of genuine interest, and to learn about the things that truly count, he needed always to stay in the city so as to be able to talk to the fellow humans he encountered there; straying outside its borders, where he presumably would come across only animals and the other elements of nature – nature as separate from and as unaffected by humans – would be a waste of time, not only philosophically speaking, but also in terms of the presuppositions for developing into a fully human individual and attaining the capacities required. As far as the historical record is concerned, it would not be putting it too strongly to say that this stance on the part of Socrates is a defining moment in Western history, inaugurating the strict ontological as well as moral divide between humans and nature, and implicitly: between humans and animals, that has for so long been its trademark – furnished, of course, with a novel explanation and justification by later giants such as Descartes and Bacon, only thereafter to be challenged by Darwin and the research done in his wake by twentieth-century scientists such as Konrad Lorenz.

Lorenz is a key figure in Midgley's work on humans and animals, and the reasons why are worth exploring. For a start, what interests Midgley is "to show how and in what cases comparison between man and other species makes sense" (1995: 23). Her point of departure is that what she calls "the Blank Paper view" is wrong, taking, as we just saw, Sartre as one of its most prominent advocates. What is wrong is the view that "man has no instincts" in the sense that there simply is no "innate determining element in human behaviour" (ibid.). Positing instinct as wholly absent in humans and as operative in all animals (to put the contrast as crudely as it is meant here) is part of the general tendency for humans to exaggerate our differences from all other species and, concomitantly, to ground all activities we value in capacities unshared by the animals. This is a deep-rooted disposition that, according to Midgley, is closely tied to the problem of evil, understood as the infliction of pain upon others. "Man", she observes, "has always been unwilling to admit his own ferocity". Unsurprisingly, here as elsewhere, animals are given a crucial function in maintaining this denial, resulting in the attempt of humans to deflect attention from our own ferocity (Lorenz speaks of aggression) by "making animals out to be more ferocious than they are" (30). This is one of many instances to follow where the irony in play should not escape us: to make the case that humans are what we are by not being like animals, references to animals are crucial.

For Midgley, the importance of the ethological studies carried out by Darwin and, in more recent times, by Lorenz, Niko Tinbergen, and Jane Goodall, has to do with the observation-based, empirical underpinning they provide to challenge the long-standing notion of human uniqueness, based in particular on the claim that rationality and language have no counterpart in other species. While granting that these are capacities exhibited to a higher degree of sophistication in human beings, Midgley shows how ethological studies demonstrate their continuity with the mental capacities of other species. Indeed, being a mammalian species, it should not be surprising, let alone contestable, that we share the general structure of motivation found in such species. There is a set of basic biological needs having to do with aggression, sex, parenting, protection of offspring, dominance and position within the group, and so on; needs that have to be met and negotiated by humans and other mammals alike, though doing so in highly distinct ways, characteristic of the species in question – whereby the interesting distinctions are not solely those between, say, humans on the one hand and chimpanzees and cats on the other, but between the latter two as well, showing up the analytic poverty of foregrounding the crude humans-animals comparison. Animals of various sorts form communities, enter into rivalries in attracting a partner; they tackle internal and external threats and distribute scarce goods to those most in need just like (or not quite like) humans do, and the abilities for social interaction in general and for anticipating the next move of a predator (or prey) are often a matter of life and death, making an intelligent use of social skills an imperative for animals all across the board. In many cases, the abilities of an animal to, say, flee from an imminent danger are far superior to that available to a human – flying is a much better response than running.

According to Midgley, one of Lorenz' most important points is made in this passage in his *On Aggression*:

> It is a widely held opinion, shared by some contemporary philosophers, that all human behavior patterns which serve the welfare of the community, as opposed to that of the individual, are dictated by specifically human rational thought. Not only is this opinion erroneous, but the very opposite is true. If it were not for a rich endowment of social instincts, man could never have risen above the animal world. All specifically human faculties, the power of speech, cultural tradition, moral responsibility, could have evolved only in a being which, before the very dawn of conceptual thinking, lived in well-organized communities. Our prehumen ancestor was indubitably as true a friend to his friend as a chimpanzee or even a dog, as tender and solicitous to the young of his community and as self-sacrificing in its defense, aeons before he developed conceptual thought and became aware of the consequences of his actions.
>
> (Lorenz 1966: 246; see Midgley 1995: 264)

If by aggression is meant the general tendency to attack, Lorenz explains, we would misconstrue it if we view it as tantamount to a tendency to destroy, as so often happens. Aggression is best understood as a *driving away*, a demand for space, be it on behalf of those belonging to one or of oneself; hence, it is a piece of behaviour that is not antisocial, but performed for social reasons within a social setting. To speak of the motives of aggression in this way applies to a number of higher-standing animals (mammals) as well as humans. The driving away Lorenz refers to is predicated on the higher development of feelings, with abilities for social responsiveness, for the intervention or response that is called for in the situation and would be appropriate to, say, the threat perceived against one of one's own. Thus understood, the aggression involved in driving that threat away springs from what may seem its opposite, namely affection (Lorenz does not use the word love); it springs from a pro-social basis, that of in-group bonds, giving rise to loyalty and the obligation to protect the vulnerable, primarily of one's own kin, but not always and exclusively. For aggression to serve its purpose, then, basic social skills and feelings need to be in place so that intelligent discriminations among the variety of others can be made, enabling a response of inclusion in the one case and one of hostility in the other, there often being a mixture or alternation between the two, not only depending on who the other is in the particular case, but also – more complexly – towards one and the same other. The capacity, even need for ambivalence we spot here, is a deep-seated one, says Lorenz, and it would be profoundly mistaken to try and root it out. Aggression, therefore, is a central element that humans share with many animals, one involved, Midgley points out, "in the development of many unquestionably valuable parts. Trying to get rid of it is something like sawing off a hand because it has held a sword" (1995: 78). The bottom line, then, that Midgley sees as the takeaway from Lorenz here, is that "like power steering in a modern car, responsible morality derives the energy which it needs to control human behaviour from the same primal powers which it was created to keep in rein. Man as a purely rational being, divested of his animal heritage of instincts, would certainly not be an angel – quite the opposite" (Lorenz 1966: 247; see Midgley 1995: 271).

Common to humans and other mammals, then, is the crucial role that social needs, social bonds, and social skills play in their life – ineluctably so, because there is no other way to survive. The fact that these social qualities are found across the human-animal distinction is what is basic; the many differences in how they are acquired and enacted are secondary, though by no means of modest importance, whereby the differences among animals are commonly downplayed as compared with those which are taken to set humans apart from all other animals.

As for the latter: what are these differences? If there is anything truly unique to the human species, what is it?

We will have plenty occasions to point out how common it has been for major philosophers within the Western tradition since Plato to answer this question by way of the *via negativa* that animals (in the unspecified plural) have been held to provide, helping bring into sharp relief what we humans find exhibited in ourselves yet would look for in vain in animals, no matter what kind.

That said, the most obvious candidate for what we are talking about here is probably "culture", taken to include such allegedly uniquely human capacities as those required for morality (including intentional evildoing, or wickedness, a topic Midgley is particularly fond of), besides practices of a sort that requires sophisticated cognitive skills, such as language and conceptual thinking of the kind involved in, say, doing philosophy. To put the notion to the test: we wouldn't be prepared to say of a cat, or a sparrow, or a trout, that it is "philosophizing", would we? Just as we wouldn't be prepared to say that such an animal, in a given situation, is behaving irresponsibly?

No, we would not. And if someone would, the charge of anthropomorphizing (more on which later) is to be expected. That surely is the stance we are culturally expected to take over this question, cultural beings that we are, unable to ask and answer any question whatsoever without the resources our culture provides for us in doing so, and doing so in the form of communication with others, at that, fellow *human* others, not others in a special-neutral sense.

"Culture is natural", contends Midgley. "We cannot live without a culture, but it never quite satisfies us" (1995: 272). Although we humans are essentially social creatures, as we saw above, we "come half-finished"; man, Midgley explains, "is innately programmed in such a way that he needs a culture to complete him" (274), and yet the completion facilitated by culture is never complete; there is always room for, nay need for, an in-between, an open space, even friction, between the individual human being and his or her culture. Culture, according to Midgley, is not an alternative or replacement for instinct, but its "outgrowth and supplement" (ibid.). Culture is essential, something we cannot live without *because* of our innate needs.

The key insight, then, is that "man is by nature a being of culture", to quote Lorenz, stressing that "man's whole system of innate activities and reactions is phylogenetically so constructed, so 'calculated' by evolution, as to *need* to be complemented by cultural tradition". Interestingly, Lorenz goes on to state that "were it possible to rear a human being of normal genetical constitution under circumstances depriving it of all cultural tradition [...] the subject of the cruel experiment would be very far from representing a reconstruction of a prehumen ancestor, as yet devoid of culture. It would be a poor cripple" (Lorenz 1966: 264f.; Midgley 1995: 285).

I find Lorenz' thought experiment interesting because it invites a parallel, one he does not mention, nor does Midgley: namely, would it be possible to rear a human being under circumstances depriving it of experiences with animals? In other words, would socialization be accomplished, would we deem it complete if present only human others, absent animal ones? I don't think this is an open question, one that we are agnostic about. Rather, I propose, we take it for granted that so-called "significant others", those crucial to the development into a distinctly human person, to the acquisition and performance of the distinctly human abilities and skills, are fellow human beings – interacting with, communicating with, *being-with* others of our species, hating or loving them, is the trajectory for humans becoming fully human and realizing their potential as such. In short, the whole process, in terms of whom to relate to, bond with, and respond to, in terms of participants in any role deemed important, is considered a one-species affair, our internal one at that.

For all its obviousness once stated, the parallel is not considered by Midgley. Why?

I suggest that the answer is more complex than we would expect. Not so for philosophical reasons, but for historical and cultural ones. Two things are being taken for granted: that humans not animals are the creatures required for human infants to evolve into fully human adults; and that the world in which this one-species process occurs, one that is started over again with each newborn individual, is a world where animals exist alongside humans – only that animals are not looked upon as crucial to the process.

But putting it like this is too abstract. The point I want to make is utterly concrete: not a fancy thought experiment but instead an observation about a rapidly changing world, one where what used to be an abundance of animals coexisting with humans is dwindling from one year to the next.

The extinction of species and the reduction of exemplars in those still left, driven by the shrinking of habitats on all continents, prompting some scholars to say that "there is no more nature, at least not in the sense of wildness", are not taken into account in the discussion we have looked at in Midgley (and in Lorenz) above. Having authored the works cited in the 1980s and 1950s, respectively, taking animals for granted is perhaps not something they should be faulted for; it is only now that doing so would be out of step with the facts as we know them.

II

What, then, is the significance of animals? Or more to the point, what would their disappearance mean?

These are tough questions to answer. For a start, since animals have been around, in abundant numbers, much longer than humans, there has simply been no basis in reality for considering the issue, not even as a relevant hypothesis. That said, the genuinely philosophical dimension should not be overshadowed by the historical aspect – namely: what would the disappearance of animals mean for whom?

To observe that the question is raised from the point of view of humans is to state the obvious. How could it be otherwise? It couldn't. *Every* question raised – be it about animals, be it about humans – is raised from the human point of view. But this does not mean that the extension of the question is confined to humans. There is something to be said about the significance *for animals* of animals being decimated in the contemporary world: they are the ones directly affected. Even if this is granted, it will immediately be observed that there is no way of saying what this significance "for animals" amounts to, since asking such a question is bound up with the human point of view. The bottom line is pinpointed in Thomas Nagel's seminal paper "What's it like to be a bat?" (1974), which is to say that in order to answer the question we would need access to the being-in-the world peculiar to the animal, an access we are prevented from ever getting because of the species-barrier we are stuck with as humans.

I shall return to the historical dimension of the ongoing decimation of animals in later chapters. Philosophically speaking, as Midgley argues in *Animals and Why They Matter?*, the issue is more complex than just suggested. The so-called "problem of other minds" has engaged several generations of the presumably most ingenious analytic philosophers, based on a deep-seated suspicion over inner experience to the effect that there is no way I can access another's experience so as to be sure about what it consists in in a given situation. The best I can do is to make a qualified guess, but in principle I may always be wrong even when that other is my wife or son. Now, if I have no foolproof or objective way of knowing what the *person* closest to me is experiencing (thinking, feeling), how can I possibly get it right when that other is an *animal*? In other words, if the access denied us, fundamentally and inevitably, concerns our attempts at understanding the inner experience of fellow humans, it is no wonder that it is denied us in the case of animals.

It is worthwhile to look closely at Midgley's account of this issue. The proponent of the sceptical position just sketched, she writes:

> is uneasy because he thinks that, if we say that a dog is alarmed or that it sees the river, we are claiming to know exactly how it feels. This would certainly be wild. But then it would also be wild in talking of human beings. Even when we know most clearly that another person is alarmed, or has seen the river, we can never share their experience. The barrier to sharing it is already a complete one with human beings, so it cannot be made any more complete by adding the species-barrier to it. We may suspect that the dog's experience would actually be more different from ours than another person's would. But that suspicion is irrelevant, since an impossibility cannot be made more impossible.
>
> (Midgley 1998: 130)

For Midgley, what she calls full-scale sceptical Behaviourism is a dead-end in insisting that we can know *nothing* of each other's subjective experiences. To stick to the example, alarm is not *only* an internal feeling; it has a public dimension in being expressed in conduct. Indeed, in the normal case inner perception and observed conduct go together, making for a correspondence that, though rough, is "good enough to give us our shared public world" (ibid.). Midgley refers to our practice of constantly checking one against the other; in trying to interpret the reactions of others – what does he feel in acting like that – we ask ourselves what our own feeling would be in that sort of situation. Besides, to invoke a point often made by Hume, owing to our membership in a community our expressive and sympathetic powers are well developed, allowing us to mutually adjust and to convey feelings to each other even though we cannot get inside each other's heads (see Hume 1969 [1740]; Baier 1991; Wiggins 1987), the point against the sceptical position being that there is no necessity that we do so.

On Midgley's view, then, the problem is not about anthropomorphism – attributing human qualities to non-human beings – but about Behaviourism in that it "arises already on the human scene". The barrier, that is, "does not fall between us and the dog. It falls between you and me" (1998: 130). But this, Midgley contends, is simply false: the powers of expression and sympathy we possess qua humans allow us to assess each other's subjective states, if not in a foolproof manner with no risk of getting it wrong, then at least, in the typical instance, in a manner conducive to successful interaction.

Getting the other's feelings roughly right is the normal; and when we do get them wrong, we will often be told as much, pointing to the importance of mutually checking on each other, which is precisely what we typically do, showing up the limits of any stance wedded to solipsism to appreciate that what we are up to in seeking to understand each other's "inner" experiences is sui generis something social, inter-subjective not intra-subjective; a practice we always already are engaged in, rather than a philosophical certainty (to allude to the shift from early to late Wittgenstein).

So far, so good. Checking on each other, making the experience or feeling in question a topic of communication, thus converting the assumedly "private" into something genuinely public, something to whose character both parties – the interlocutor as well as the I – contribute, helps drive the final nail into the coffin of scepticism. Midgley goes on to ask: does this work with other species? That is to say, "do human beings, in their frequent interactions with other animals, manage to identify their moods and feelings too?" (1998: 133).

Midgley's answer is pragmatic. The question, she tells us, can only be answered "in terms of success, of improved interaction, resting on interpretation and prediction" of the animals' actions. She adds that this is true on the human scene as well and that "those who try to understand animals, and give time and attention to the matter, often come to understand them quite well. Those who do not, fail, which is also true with human beings" (ibid.).

There are two ways to respond to this. The first one is that Midgley's pragmatism is to be welcomed: she brings the philosophical debate over the species-barrier down to earth by demystifying the alleged radical "otherness" of animals, suggesting that, in principle, *they are not more difficult to understand than our fellow humans*. The other response is that in putting the emphasis on how similar the two cases are, she misses the differences and so dodges the philosophical challenge they pose.

But what precisely does this challenge consist in? Midgley makes it clear that the burden of proof does not rest with those who believe in subjective states in animals; it rests with those who deny it. Subjectivity is not "an extravagant metaphysical hypothesis" (1998: 134); this holds for animals no less than for humans. All we need to do is to start from the facts to be explained, like this: "Granted the similarities between nervous systems, and also between many of the ways in which animals interact and the ways in which humans do, and granted also that humans often interact with animals successfully, *how wide a difference in their basic social faculties does it make sense to posit?*" (ibid.). In other words, we should not posit a greater difference between the causes than is needed to suit the differences in the effects; pragmatism á la Midgley here means simply to be economical. Again, the argument against scepticism is not theoretical; it is instead eminently practical, a fact about our own experiences with animals that we cannot deny: "Every day and all around us, even in our urban and restricted way of living, people rightly assess the moods of dogs, and dogs of people.

Cats, horses etc. are also doing very nicely. In the Amazonian jungle, where it really matters, they do better still" (ibid.).

Notice how effortlessly Midgley moves from talking about how humans understand animals to how they understand us. She evidently sees no more reason to cast doubt on the fact of the latter than on the former. Much as I find myself in general agreement with Midgley's critique of scepticism, it is one thing to challenge the basis for a "deep divide" between the species, and quite another to replace it with the sort of symmetry that Midgley seems to be proposing here, without really arguing the case for it. That cats and horses are able to assess the moods of humans can be easily granted, I take it, bearing in mind that the humans in question will be the selected ones with whom such animals – the cat being a pet, the horse a domestic animal – interact on a daily basis and often throughout their lives, so that should they get their people wrong, they will be corrected, and hence enabled to get it right the next time around, with that particular person. But what justifies the claim that the animals in the Amazonian jungle – mostly non-domestic ones – exhibit this capacity "better still"? What one can say is that their lives depend on getting the humans – say, hunters – right; nothing less than survival is at stake, and this goes both ways, applying to the humans' need to get it right as well. To observe as much is merely to stress what is at stake; it is not to show what makes it possible.

Midgley reminds her readers of how much has happened since philosophers such as Descartes and Kant wrote about animals; in fact, "the last half-century has seen a complete revolution in our understanding of animal motivation, and still more of animal communication" (1998: 136). Traditional knowledge has always existed, yet, in spite of Darwin's path-breaking research, it was largely forgotten. Only in the twentieth century did zoologists help inaugurate a shift in the understanding of animals, writes Midgley; ethology was established as the scientific study of animal behaviour, and Lorenz's work has been particularly influential. Again, according to Midgley, this research is to be regarded as in continuity with the study of humans and not as a radical departure from it. Hence the pioneers in ethology "found it entirely natural and necessary to use the regular vocabulary already employed to describe human motive. This (as I have suggested) is not surprising, seeing that it had in any case been invented for use on animals as well as humans" (ibid.). But whereas Lorenz does not hesitate to talk about the subjectivity of animals, his colleague Niko Tinbergen is not quite so sure: "Because subjective phenomena cannot be observed objectively in animals, it is idle either to claim or deny their existence" (quoted in Midgley 1998: 137). But if one holds, as does Tinbergen, that feelings are a sort of phenomenon that can be known only by introspection, his argument must apply just as much to people as it does to other animals, which amounts to the solipsist position that Midgley rejects.

On this basis, the notion that anthropomorphizing – understood as the attribution of experience (or more precisely: of the capacity for experience) to other than human beings – is detrimental to science and must be rooted out, is deeply mistaken according to Midgley. The attribution of emotions to animals, she says, is just as justified as are our beliefs about the inner states of other people – both versions of this practice of ours have proved generally successful (see 1998: 142), and no further proof is needed. Of course, a distinction must be made between holding that animals *have* subjectivity, that they have feelings such as fear or joy, taken to be of the same general kind as those we humans have, on the one hand, and getting it right about what kind of feeling that animal is expressing in a particular situation, on the other.

As to the issue of the subjectivity of animals, there is reason to question the correctness of Midgley's claim about a strong continuity from Darwin to ethologists like Lorenz and Tinbergen. As Eileeen Crist shows in her study *Images of Animals: Anthropomorphism and Animal Mind*, Darwin's portraits of particular animals and their behaviour employ anthropomorphic vocabulary throughout, non-metaphorically so.

In turning to the works of Lorenz and Tinbergen, readers will be struck by a striking contrast in that, if at all using mental predicates such as love, anger, jealousy, grief, pretending, and so forth, these concepts will typically be in scare quotes, a practice considered as required to come across as giving a truly scientific account of the animals and so functioning as a marker of scepticism, as if unqualified and unequivocal commitment to the reality of "animal mind" would violate the requirements of rigorous research. As Crist demonstrates by way of numerous textual examples, ethologists depart from their naturalist predecessors in that their accounts of animal behaviour patterns are generic: they are "accounts of the 'typical' case, not the singular instance", not a particular episode involving a particular animal whose differences from others – hence personality – fascinate the scholar like it did Darwin (Crist 1999: 74). Referring to Tinbergen's work on the carnivorous water beetle, Crist lets us see that according to ethological reasoning "it is not even that animals react to stimuli, but rather that key stimuli release their reactions", so that, to quote Tinbergen, "the beetle's feeding response is *released* by chemical and tactile stimuli exclusively; for instance, a watery meat extract promptly *forces it* to hunt and to capture every solid object it touches" (Tinbergen quoted in Crist 1999: 109). The overarching claim is that the stimuli work as *causes* of animals' responses, advancing a mechanistic picture of the animal and a rigid image of animal behaviour, portraying animals' activities as "non-voluntary, as causes act on or through the animal, bringing about what the animal does as something that happens to it", hence casting it as essentially passive not active, as reacting rather than as initiating action (1999: 122).

To anticipate a point to be elaborated more fully below, the approach of the ethologists is testimony to the primacy of method over subject matter. Crist explains what is at issue by way of the contrast invoked above: "The naturalist goes into the field to record the stream of life as it occurs"; he or she "seeks to record singular events witnessed in great detail", and with a keen and curious eye for the ways in which the distinct subjectivities (personalities) of the animals are played out in their actions. By contrast, the ethologist (or sociobiologist) "goes to the field to collect the data that are required by the hypothesis formulated and the experimental design" (1999: 146). The result is that in the ethologists' research, individuals become absolute incumbents of (theoretical, predesigned) categories, so that "descriptions of animals' activities are always generalized, delivered through quantification markers such as *often, frequently, regularly, rarely*, and the like". As a systematic consequence, "the idiosyncrasies and particularities of individuals, the ways in which their uniqueness is expressed, are factored into the analyses only indirectly to the extent that they are pooled into quantitative operations. In a discourse that is fundamentally statistical in character" – in keeping with the primacy of method, of experimental design – "individual variability, as such, is minimally significant" (1999: 147). Again, the use of scare quotes in depicting what would be the equivalent to human motives, to emotions, thoughts, and projects in animals and as exhibited in their behaviour is considered mandatory for any ethologist that aims to come across as doing proper science, dismissing as anthropomorphic any vocabulary that violates the border between human and animal worlds, human and animal minds – granting that there is such a thing as the latter, which is exactly what the scepticism Crist demonstrates is dressed up as "hard science" disallows. Whereas the distinction between behaviour and action are largely interchangeable in ordinary usage, Crist points out how their essentialist bifurcation as carried out in the ethological research she reviews "guarantees that animal and human affairs remain qualitatively separate", thereby calling for and justifying "discrepant frameworks of description and explanation" (1999: 210). Regrettably, Crist does not explore the wider societal and cultural consequences of the separation she critiques within ethological research; in fact, only in her study's final sentence does she point to this wider dimension, observing that "the hierarchical regime of higher-to-lower is ready-made and often unquestioned, and, in setting humans apart from the animals, it has fuelled our self-importance and propped our thoughtless and destructive relationship with the natural world" (1999: 222).

To be sure, there is reason to question whether the separation between human and animal minds, to the point of dismissing subjectivity in animals as "unscientific" and "speculative", as the product of anthropomorphism remains as predominant within the scientific community today as it was some decades ago. Crist acknowledges that her material mainly consists of the work carried out by Lorenz and Tinbergen in the 1950s and 1960s. But she argues that Edward O. Wilson's acclaimed *Sociobiology: A New Synthesis* (1975) perpetuates the same sort of approach, philosophically speaking: "Sociobiology", she writes, "is conducive to the sort of elision of mentality referred to as 'methodological behaviorism' in philosophical terminology. Mental tropes are not explicitly denied...but simply assigned no work in the production of behavior" (1999: 228, n. 13).

As for Midgley, one looks in vain in her work for a critique of ethology along the lines carried out by Crist, which is surprising given Midgley's lifelong interest in issues within the philosophy of science. The variant of scientism that Midgley never tires of attacking, for what it's worth, is the one she encounters in Richard Dawkins' *The Selfish Gene*, whereby the problem she sees is to do with the misrepresentation and distortion of what is specific to *humans*, not what is specific to non-human animals (see Midgley 2001, 2010, 2018).

III

Midgley ends *Animals and Why They Matter* by saying that the book has been occupied with "a negative", with "removing barriers which our tradition has erected against concern for animals" (1998: 144). Creatures like crocodiles and dandelions, she remarks, tend to be rather unpopular with the human race. But there are three sorts of human beings who beg to differ, who see the point in their existence and befriend them – namely poets, scientists, and children (145).

To be sure, this is an odd group, comprising what Midgley elsewhere – most notably in *Science and Poetry* – argues are two ways of looking at the world that often have been at loggerheads. As for children and their characteristic openness to animals of different kinds, more below.

Now if Midgley is right that our tradition has erected barriers against concern for animals, why is that so? And why have those barriers proved so hard to remove?

Midgley credits Darwin's theory of evolution with the advance of science that connected human beings more firmly than ever with the other animals. The case was made, by systematic observation in keeping with the requirements for modern scientific rigor, for a strong continuity between human nature and the nature of other creatures. Thus, Darwin set about to raid the full range of psychological concepts that have been developed for describing human feeling and behaviour; the more varied and richer is our understanding of these traits in ourselves, the more can we expect to be able to describe the traits we share with other animals. If there is no barrier but instead *continuity* between humans and (other) animals, comparison not only makes sense but becomes the principal way of getting at significant differences as well as similarities between humans and non-humans, whereby the differences are thought of as gradual and in terms of degree rather than absolute and static.

In *The Myths We Live By*, written much later than her two early volumes on humans and animals, Midgley acknowledges that after Darwin's death, "the tide turned against all such thinking" (2004: 140), by which she means to include – problematically so, given Crist's above account – the pioneers of modern ethology, Lorenz and Tinbergen. Why?

On this occasion like on earlier ones, Midgley is quick to point to the role of Behaviourism: its peculiar psychology did indeed "officially treat humans and other animals as similar, but it did it in exactly the opposite way of Darwin, by treating both as insensate machines" (ibid.). While John Watson's heirs felt compelled to abandon the view that there was no such thing as a thinking subject (not even in the case of humans, that is) and that "consciousness" was merely a superstition, academic psychologists have nevertheless, writes Midgley, continued to use mechanical models for the study of humans, alongside "nominal admissions that subjectivity was present", and leaving the resulting tension unresolved. As for animal psychology, "pure mechanism still largely ruled, because it was still held up as 'scientific'" (ibid.). Sociologist and anthropologists also did their part to uphold this situation in continuing to treat human beings as unique, denying – pre-Darwin-like – that comparisons from other species could be relevant to the social sciences. Indeed, the presence of inborn behavioural tendencies in humans was just as anathema to sociologists as to behaviourists, according to Midgley, who cites the professionalization of science as an important factor in the change from Darwin's position involved here.

On my view, there is something dated about the pride of place Midgley accords to Behaviourism. Obviously, Darwin was recognized as a scientific giant in his own time, following the publication of *On the Origin of Species* in 1859 and the wide-ranging reception it was met with, not unanimously favourable, of course, but unequivocal in that a major paradigm shift was being inaugurated and widely recognized as such. So my question remains: why was there a significant step backwards in the wake of Darwin? Why was the notion of human exceptionalism so stubbornly held on to? Why was there still a career in academia for the mechanistic view of animals propagated by Descartes' "mere automata" thesis, predating Darwin by some 200 years? There must be something more fundamental at work here than the professionalization of science.

More promising than Behaviourism in this respect is the other powerful twentieth-century philosophy that Midgley keeps referring to, namely Existentialism. Of course, Existentialism as worked out by early Sartre, its main protagonist in Midgley's narrative, can have no track with Darwin's insights into human-animal continuity. There is no such thing as human nature; for humans, existence precedes essence, much as *the made* – artefacts, culture in general, as the ongoing creation of humans – outshines *the grown* – animals and plants, the natural world in general – in terms of importance for everything of human and philosophical interest alike. The divide between human and animal, mind and matter, freedom and necessity, is just as absolute and unbridgeable in Sartre as in Descartes, notwithstanding the three centuries of immensely influential scientific breakthroughs. The whole point about the human project, according to Sartre, is to transcend nature, understood as the domain of inertia, comprised of entities that simple are, and will forever remain, what they are, self-identical throughout time, and as such in stark opposition to humans who are what they are not and not what they (at any particular point in time) are, owing to their unique ability to "negate" and project themselves beyond any given state of affairs. Given this outlook, to think that there is anything that I as a human share, ontologically, with the natural world and all the mindless, purposeless creatures encountered there, anything at all that would show us up as on the same footing in some important respect, say, in terms of (inter) dependence, is tantamount to engaging in bad faith and so in self-deception. This, I suggest, is *anthropomorphism in reverse*: attributing to oneself qua human a sort of inertia found in nature, something static and immutable, precluding dynamics and change, meant to justify that "I had no choice", this is just the way I am, and things generally are, thus turning myself into a thing-like object and renouncing my status as a free subject responsible for everything I do – this total freedom of mine being the one given condition of my existence I cannot undo.

It follows that Existentialism á la Sartre is totally and unashamedly anthropocentric, not only descriptively so but normatively as well: only the things that result from our activities, exhibiting our powers to alter any given state of affairs, have true merit and satisfy our craving for meaning and purpose, qualities that the natural world, in part and in whole, is devoid of. In *Beast and Man*, Midgley's critique of this philosophy is nothing less than devastating, pointing out that "the really monstrous thing about Existentialism is its proceeding as if the world contained only dead matter (things) on the one hand and fully rational, educated, adult human beings on the other – as if there were no other life-forms" (1995: 316). We have to do with an expression of "desertion and abandonment" which, Midgley suggests, is due not to the removal of God (Nietzsche), but to a "contemptuous dismissal of almost the whole biosphere – plants, animals, and children" (Evernden 1993: 119). This harsh criticism recalls that put forward by Hans Jonas, a thinker congenial to many of Midgley's concerns yet nowhere mentioned in her work. "No philosophy", writes Jonas, "has ever been less concerned about nature than Existentialism, for which it has no dignity left" (1966: 232). For Jonas, the existentialist depreciation of nature that he finds in early Heidegger as well as in Sartre must be seen as reflecting nature's "spiritual denudation at the hands of physical science". The metaphysical "given" in Sartre, that only humans care and that nature does not care, that nature is devoid of all "primary" qualities such as volition and purpose, intelligence and value, experience and sentience, and as such indifferent, is the true abyss according to Jonas.

What is less than fully explained in Midgley, however, is why a philosophy so openly contemptuous of nature could become so popular in post-war Western society and remain so for many decades. Why did this way of thinking about the relationship between humans and the rest of nature resonate with so many people? And how could a major philosopher like Sartre get away with championing such a crude – either/or, all or nothing – portrait of this relationship? Why didn't he even bother to make distinctions within that vast domain "nature", doing justice to the difference between animals, plants and stones, lumping the organic and inorganic together under one umbrella and pretending that they are equally devoid of the properties of intelligence and feeling, purpose and direction, that humans are said to be the sole bearer of? Why didn't this outlook strike people as deeply counter-intuitive and at loggerheads with the way they experience and treat different creatures in nature?

Perhaps this is more a question for the cultural and social sciences than for Midgley. And yet, there is more to it, for what is striking once the above questions are raised is not only that Existentialism of the sort Midgley attacks seems hopelessly dated but that there is something dated with Midgley's own position as well.

IV

I mentioned that Midgley pays special attention to the way in which children and animals interact. "It is striking", she writes, "how domestic animals will put up with rough treatment from small children, which they would certainly not tolerate from adults" (1998: 117). She goes on to make a more general point: "It is not just that most human beings have in fact been acquainted with other creatures early in life, and have therefore received some non-human imprinting. It is also that children who are not offered this experience often actively seek it. Animals, like song and dance, are an innate taste" (118).

These are not only strong statements of a type never found in Sartre, for whom children, just like adults, will do perfectly well without animals and have no need for their presence so as to become fully human. More to the point, Midgley offers them in a timeless manner, as expressing observations both evidently true and as borne out in practice in ways familiar and available to all. Knowing full well that the relationship between humans and animals is eminently historical as well as cultural and so subject to major shifts, Midgley nonetheless is guilty of the philosopher's predilection for treating her topic *sub specie aeternitatis*.

Recent years have witnessed a plethora of novel concepts coined in order to come to grips with the ways in which the natural world is undergoing changes never seen before. With regard to Midgley's work, the most relevant such concepts are "nature-deficit disorder", "Solastalgia", and "extinction of experience", coined by Richard Louv, Glenn Albrecht, and Robert Pyle, respectively. Their common concern is the fact that fewer children connect with nature in the way of first-person, bodily sensuous experiences of, and encounters with, animals that are neither pets (cats and dogs) nor domestic (horses, hens, pigs, and cattle). The less wilderness there is left, the more the habitats of these animals shrink and become fragmented, the fewer will be the arenas where such encounters can occur. That this change is happening, at increasing pace, tends to go largely unnoticed, by children and philosophers alike. Two academic terms for this phenomenon have been coined, namely "shifting baseline syndrome" and "environmental generational amnesia". Psychologist Peter Kahn explains:

> We all take the natural environment we encounter during childhood as the norm against which we measure environmental degradation later in our lives. With each ensuing generation, the amount of environmental degradation increases, but each generation in its youth takes that degraded condition as the nondegraded condition – as the normal experience.
>
> (Kahn quoted in Peterson 2009: 92)

This means that what is objectively a dynamic situation of decline and degradation in our natural surroundings, even those most familiar to us by first-hand experience, is for the newcomer to the scene, the child, normality as such, the only situation known, and with that, nothing to be concerned with let alone alarmed at. What shifting baseline syndrome amounts to is a displacement of the interpretive criteria for what is perceived as alarming in a sense involving changes that might prove dangerous – a displacement that is objectively effective (affecting people's understanding of reality) without being subjectively recognized *as* displacement (see Lertzman 2015). This is a subtle psychological phenomenon whose effects are eminently concrete: children who grow up with little or no snow will not be in a position to miss the winter-related plays (building a snowman) and activities (sledging, skiing, skating) that their parents took for granted, even when growing up in the same place. This goes for the perception of animals as well: their gradual disappearance from the scene, occurring from one generation to the next, goes largely unnoticed, indeed tends to be totally lost on the youngest generation. The question is: Can you miss – notice as missing, lament as missing – something you have never experienced?

I take it that this is a phenomenon that can be noticed in many families. The young British writer Lucy Jones relates how she came to realize that her grandmother, in virtue of the state of the natural world she was born into, had acquired "an inherent lexicon of that world and how it operates". Likewise, Jones goes on, "my parents knew their birds, flowers and plants; names, timing and behaviors. I knew a bit, maybe 5 to 10 per cent of what they knew" (2020: 8). No wonder, then, that Jones, upon having her first child, asks herself if her daughter will be able to name – by which she means know – anything at all. The increasingly fragile experiential connection between humans and their natural world that we have started to notice involves, in the words of the ecologist Robert Pyle, "a cycle of disaffection and loss that begins with the extinction of hitherto common species, events, and flavors in our own immediate surrounds; this loss leads to ignorance of variety and nuance, thence to alienation, apathy, an absence of caring, and ultimately to further extinction" (Pyle quoted in Jones 2020: 8).

Against this background, we are in a different position than earlier philosophers when it comes to acknowledging the connection between human experience and the natural world: losses to the point of extinction (of species, of habitats) in the latter leads to a shrinking, ultimately extinction, of the former. As a matter of historical fact, virtually everything that a typical child growing up in a country like Norway or Britain today will experience first-hand – be it touch, see, hear, taste or smell – will be artefacts, things made by humans and to be used by humans. In other words, children will be relating to something *made* rather than to something *grown*; they will engage with others within the domain of culture rather than with others within the domain of nature (if that crude distinction can be tolerated here), hence with fellow humans rather than with non-human such, be it animals that are not pets. The big picture is that it is only as recently as since 2007 that more than 50 per cent of the world's population live in cities; by 2050, 68 per cent will live in urban areas. As of 2018, the figures are 82 per cent in Northern America and 74 per cent in Europe (see Jones 2020: 183).

These are well-known trends often commented on. But their role in effecting a major change in the relationship between humans and animals has yet to be fully explored. "No animal", writes Midgley, "is just a simplified human being, nor do children take them to be so. However friendly they may be, their life is radically foreign, and it is just that foreignness which attracts a child. The point about them is that they are different" (1998: 118). Yet holding this does not prevent Midgley from warning against the "chronic, endemic exaggeration of the differences between our own species and others" she finds widespread, blaming biologists as well as dogmatism in the social sciences (2004: 141). Perhaps we should see no tension, let alone contradiction between the two statements cited: appreciating factual difference in one-to-one encounters is what we do in practice, exaggerating that difference so as to maintain a cultural species barrier is what we have long tended to do in theory. An animal in its otherness from us is what attracts us spontaneously, as borne out in children's innate fascination with how differently from us a given animal will behave, exemplarily exhibited in the manifoldness of movement and of sound, showing up how many ways there are, say, to signal attack or to display fear, motives, and feelings common to us and them, indubitably so in real-life encounters, notwithstanding philosophers who, armchair-style, insist on the fallacy of relying upon what our senses disclose to us about those others.

As far as a scientific approach is concerned, are animals subjects or objects? Do they have to be made into objects to qualify for study? "Animal", observes the philosopher Neil Evernden in *The Natural Alien*, "is, by convention, the name of a thing, an object, a clever machine" (1993: 77). It is because we, as a culture, are not prepared to treat animals as subjects that we continue to be offended by comparisons between ourselves and other animals; animal-like means thing-like, primitive, a mere object, passively awaiting whatever use we humans intend to put it – yes, it – to, a means to suit the ends of a subject, a he or she that can only be a human such.

If humans are offended by being compared to an animal ("you behave like a pig"), what about the insult to the animal in question? A consequence of the cultural taboo against anthropomorphism that is inseparable from Galileo's and Descartes' definition of scientific method, observes Evernden, is the encouragement of total ignorance of animals – a practice, he adds, "religiously followed by many humanists". If that sounds dated in the present era of the Anthropocene, of new materialism and of posthumanism á la Bruno Latour, Evernden's principal point is not. "Science", he continues, "also helps perpetrate such ignorance, paradoxically, because it gives us a mechanistic image instead of experience of the animal. So long as we have limited direct experience of the creatures in question" – and I have argued above that the arenas for one-to-one encounters are dwindling at record pace – "it is relatively easy to accept a cultural stereotype in its place, in this case animal-as-object" (ibid.). The critical point – the viewpoint from which being critical becomes possible – is personal experience, because what Evernden calls "a much more difficult substitution" must take place for what is deemed serious scientific study of animals to take place: "The student who enters a biology program with experience of animals as subjects must 'unlearn' his initial understanding before he can proceed to a biology of objects" (1993: 78); in particular, such a student needs to unlearn his "unscientific" propensity to relate to the animals sample in the lab through empathy, being a faculty out of place here owing to its alleged subjectivity and so unreliability, revealing how it is that not only the animal object of study, but the humans carrying it out as well are required to disavow subjectivity for objectivity's sake. "How", asks Evernden, "can we permit this reversal of the primary and the secondary, our own experience of the world and an abstraction about it which for most of us really amounts to second-hand information" (ibid.)? This is the sense in which poets such as Blake and Wordsworth saw the rise in science as coinciding with, indeed as demanding, a drastic diminution in human perception, allowing only a limited range of experience to qualify as "real" and so as compatible with science, whereby experience qua emphatically that of a particular individual with a particular history encountering a particular animal was tabooed.

As Alfred North Whitehead contends, the effort made by Descartes and his successors Locke and Newton to replace nature as "subjectively" accessed by means of full bodily sensuously experience, taking in the fullness of taste, smell, and colour – of life being lived – encountered in the natural world, say, in that particular owl peeking out from that particular tree, by "nature as a dull affair, soundless, scentless, colourless; nature as merely the hurrying of material, endlessly, meaninglessly", has proved to be "an enormously successful operation". In fact, Evernden argues, Whiteheads observation in 1925 still holds: the view that attained prominence in academia in the seventeenth century, "is still reigning. Every university in the world organizes itself in accordance with it. No alternative system of organizing the pursuit of scientific truth has been suggested. It is not only reigning, but it is without rival. And yet – it is quite unbelievable. This conception of the universe is surely framed in terms of high abstractions, and the paradox only arises because we have mistaken our abstraction for concrete realities" (Whitehead 1967 [1925]: 54f.; Evernden 1993: 53). We have, that is to say, become entangled in what Whitehead famously identifies as "the fallacy of misplaced concreteness", mistakenly thinking that in order to get at what is real in the natural world, we need to relinquish subjective experience and replace it by abstraction, thus looking away from the very qualities that makes the entity what it is, say, an eagle owl.

No doubt many biologists, zoologists, and botanists would rush to have their protests registered. It is tempting to say: rightly so. But is it that simple? Has the rival that Whitehead said was missing when he lectured at Harvard almost 100 years ago, come to prevail over the largely Cartesian model of (natural) science?

Of course, this is a question so big and consequential that I cannot even begin to answer it here – but I shall return to it later. Caroline Merchant has the following to say in the Epilogue to her seminal study *The Death of Nature: Women, Ecology, and the Scientific Revolution*, first published in 1980: "The mechanistic view of nature, developed by the seventeenth-century natural philosophers and based on a Western mathematical tradition going back to Plato, is still dominant in science today. This view assumes that nature can be divided into parts and that the parts can be rearranged to create other species of being. … Mathematical formalism provides the criterion for rationality and certainty" (1990: 290). In fact, Merchant repeats her overall assessment in her most recent book, *Autonomous Nature*: "Mechanistic explanations are still the norm in molecular biology, genetics, neuroscience, parts of biochemistry, chemistry and physics, and parts of engineering and biophysics" (2016: 150), adding however that the paradigm is increasingly challenged by chaos and complexity theories, as well as by quantum theory (see Barad 2007; Wendt 2015).

In my view, the taboo against anthropomorphism Midgley refers to is even more important than she realizes, even today, having become so deeply entrenched in our culture as to part if its second nature, a key component of a mentality passed on from one generation to the next, be it among lay people or among scientists. As indicated above, this is to do with the double banning of subjectivity – comprising both the human and the animal – to attain and secure objectivity in the sense required for genuine knowledge about animals and their behaviour. Anthropomorphism, understood as a deep-seated feature of human subjectivity in its "pre-scientific" state, was banned from the laboratory, from observations, experiments, and vivisections of animals because it was seen as incompatible with the new science inaugurated in the Renaissance. Only by removing all manifestations of the human subject's seemingly innate inclination to project his own felt aliveness, his state of mind or feeling, onto the physical reality and the data encountered there, could science proper get to work. What we need to recognize more clearly than Midgley does is that this amounts to an enormous restriction – impoverishment – of human perception and experience of the natural world, one required by institutional design – by way of the socialization of new generations of scientists – so as to achieve the compliance sought; that is to say, a homogenization of the subjects and the capacities they were allowed to exercise that, in turn, helped constitute what amounts to an ongoing homogenization of their selected object domain: the same looking for the same. Following from this operation, only properties allowing for measurement, for comparison, calculation, prediction, and control within a wholly quantifiable framework qualified as apt for the requirements of the "exact" knowledge now attainable. Jonas credits Kepler with being the first among the moderns to "declare quantity the, at once, essential and truly knowable aspect of reality" (1966: 67). Accordingly, cognition consisted in measurement and comparison of measurements within a standardized setting permitting replicability of observations and repeatability of the experiments, making sure that there would be no deviance in the results obtained from one case to the next, the one scientist/observer/experimenter being eminently interchangeable with the other inasmuch as their distinct subjectivities were successfully neutralized. In this way, *method gained primacy over subject-matter, epistemology over ontology*, recalling Crist's critique of ethology reviewed above. The formal-mathematical, abstract, and universally applicable requirement of measurability came to define, indeed to exhaust, the domain of the knowable, banning the qualitative aspects of life (Locke's "primary sense qualities") in its manifoldness so completely as to render *the lifeless* definitional of what could be strictly known, and so as tantamount to *reality as such*.

To be sure, this is an old story told many times and with a strong critical edge, though typically by mavericks such as Whitehead and Jonas. When I asked above whether the self-understanding of natural science still conforms to the philosophical foundations and the requirements of method just sketched, I observed that the answer will depend on which of the natural sciences one has in mind. Relevant here is also the rise of ecology as distinct science in its own right, coined by Ernst Haeckel in 1866 as the holistic study of how organisms interact with each other and with their wider environment, whereby the flourishing of one organism (or species) depends on that of the others, thus rejecting the notion of self-sufficiency on the elementary level of each separate part (entity) that had been the hallmark of atomism and mechanism á la Hobbes (see Callicott 1989). Put in the parlance associated with Darwin, collaboration in terms of inter-species inter-dependencies is the most basic law of life, more basic even than competition – an insight that Midgley never tires of telling her readers is fully recognized by Darwin himself, who would be the first to protest his reputation as a champion of dog-eat-dog competition as what long-term survival requires and selfishness as what morality springs from as well as first and foremost has to contend with (see Midgley 2010).

My question can now be raised in a more specific manner: If we grant, on the basis of what may appear to be overwhelming evidence, that the paradigm of mechanism, of nature as dead matter, and of animals as split from us humans by way of an "ontological" species barrier, is compromised and has given way to robust alternatives of a wholly different philosophical persuasion, why does Midgley say the things she does? Why, to return to my point of departure, does she think it necessary to point out that "animals are not machines", that "actually only machines are machines" (1995: xxxvi)? Who needs to be told? Present-day scientists? Or the non-academic public? Likewise, consider what she writes in one of her latest books, *Are You an Illusion*, published in 2014, where she identifies what she takes to be "two current strange assumptions", namely, first, "that we are originally quite separate from each other", and, second, "that the direct deliverances of our own experience are worthless until they have received official scientific authorization" (2014: 54). Why does she expect it to be "surprising" for her readers to learn that "it is an objective fact in the world that our own experiences – the subjective sources of thought – are every bit as necessary for it as the objective ones such as brain cells" (56)? Why, to pick a final example, is it necessary to observe that "the idea that the world consists of objects without subjects [understood as creatures with needs, tendencies and directions of their own] does not really make sense at all" (62)?

Has time really stood as still as Midgley seems to suggest? If she is right that these are so many "myths we live by", to this day, why is that so? And where are we to look for the societal agents, or institutions, most instrumental in our continuing to live by them as a culture?

V

In Paul Shepard's book *The Others: How Animals Made Us Human*, he mentions Midgley's *Animals and Why They Matter* in a footnote, commending it as "perhaps the most thoughtful essay on the question of an ecological ethic", only to add that "one looks in vain there for the answer to the question: Why do they matter?" (1997: 354, fn. 5).

I agree. Midgley identifies and discusses an impressive number of the key questions to do with animals in the contemporary world, but she fails to answer the most fundamental one captured in her title.

As signalled by its title, Shepard's book sets out to appreciate animals as "Others in a world where otherness of all kinds is in danger, and in which otherness is essential to the discovery of the true self" (1997: 5). Hence the concept of ontogeny is the key to our relationships to nature; ontogeny understood as a kind of "necessary pattern of growth toward maturity in which we acquire respect for that which is unbridgeable between ourselves and the animals" (ibid.). Shepard finds what he is after in primal peoples and in modern naturalists alike: an attitude of "accepted separateness", one that he also sees epitomized in the work of such diverse scholars as Charles Darwin, Konrad Lorenz, Adolf Portmann, and Edward O. Wilson. What attracts him, he explains, is a kind of humility that he contrasts sharply with "civilized kindness"; a humility born of curiosity and instilling a respect for the power of animals, respect of a sort Shepard views as undermined by the culturally predominant stance of "mercy" towards animals, which he condemns as "dangerous and anemic" (ibid.).

As the reader gathers on page one, Shepard is not a thinker shy of conflict and antagonism; there is little in the way of British understatement. Shepard was born in 1926, six years later than Midgley, and he studied wildlife conservation in Missouri when he came across Aldo Leopold's soon-to-be-classic *A Sand County Almanac*, published in 1949, shortly after Leopold died. That book, Shepard tells his readers at the start of what was to be his own last work, framed the question that haunted him for the rest of his life.

What question? The contradiction of both loving and killing animals compels us to ask whether it is one that must be resolved, overcome, or one that should be held on to, one whose persistence is not only to be tolerated but recognized as invaluable. To love animals and to kill them, to propagate a land ethics and to speak, in the same book, about "the joy of seeing the kicking red legs of a shot duck dying in the morning sun" (ibid.), which is what Leopold did – this is the tension that for most of recorded history characterizes the human-animal relationship and that Shepard urges should still do so.

By placing the loving-killing contradiction at the centre of his inquiry, Shepard's approach differs conspicuously from Midgley's. For her, the question that has dominated the discourse since Peter Singer published *Animal Liberation* in 1976, namely whether it is ever morally permitted to kill animals, is by no means the most important one – coming to truly understand, and appreciate, animals in their differences as well as similarities from us, is. Hers is a project of exploration so as to get the facts right, doing justice to animals and thereby, hopefully, putting an end to misunderstanding them, something that is just as important to her as stopping mistreating them, whereby the two, perception and conduct, are intimately connected. The way I read her, for Midgley the never-ending intellectual effort to come up with an unequivocal answer – a yes or a no, once and for all – to the question of killing animals has been a great misfortune.

To put it like this is to recognize an affinity between the two philosophers' projects that we would not expect, given their different points of departure. Common to both is also a keen interest in the wider contexts – cultural, social, and moral – which help frame and negotiate every concrete human-animal encounter; an attention to the role played by historically changing settings that is often missing in the contributions of analytic philosophers. *Experience* not theorizing is the key that we as humans *and* philosophers have to animals and what is important about them being what they are, not us – and the experienced fact that they are not (like) us does not detract from their importance, does not spell inferiority on their part.

There is one more thing to note about Shepard by way of introduction, however, lest we miss his originality: his emphasis on the *cosmological* significance of animals for the evolution of human thought and imagination, for the distinctiveness of the human repertoire as a symbolic being, a repertoire that, for all its differences from that exhibited by animals, can only become fully human by virtue of exposure to animals and so precisely not in splendid isolation from them. Becoming human, then, is not a solely human process but one relying on coexisting with animals, and not simply as our point origin in some remote evolutionary past but is a major presence in our present life in the form of experiencing animals first-hand and relating to them directly, as so many individuals. The one big idea in Shepard's work is that without animals, we would never become human. Conversely, to lose the animals would be to lose the very condition on which our effort to acquire humanness depends. To become what we are supposed to be is facilitated by the kinds of beings that we are not and that for their part resist, and have a right to resist, being us. To be sure, this echoes Hegel's thesis about "Anerkennung", about "Sich-erkennen-im-Anderssein" (see Hegel 1977 [1807]; Williams 1997), only now the Other indispensable to that process of becoming what one is meant to be is an animal other, not a human one.

It is crucial that we appreciate how Shepard's approach differs from the ethological one pioneered by the figures Midgley always turns to (Lorenz, Tinbergen, and Goodall) as well as from the kind of research carried out today by scholars like Michael Tomasello and discussed by philosophers like Jürgen Habermas (2012, 2019). This means that *communication* is not the major issue for Shepard, understood as the question about how sophisticated capacities for representation we may find in higher-standing mammals such as chimpanzees or elephants, their demonstrated likeness to humans in experiencing emotional states such as grief over a lost loved one, their ability to engage in role-playing with humans, to understand the logic of taking turns, of counting, of make belief, of humour, etc. All of this is certainly fascinating, but it does not capture why animals matter from Shepard's point of view.

So what is this mattering essentially about? Characteristically, Shepard starts by asking why it is that poetry finds animals irresistible, why there would perhaps by no such – presumably uniquely human – practice without them. He gives the following explanation:

> The meaning of animals is implicit in what they do: eat, run, leap, crawl, display, call, fly, mate, fight, sing, swim, hide, slither, climb, and die. One or another does each of these things with more finesse and more expertise than people. Their keenness is reflected is the shapes of their

> bodies and the traces they leave. Animate signs and signatures move through our dreams and imagination, evoke our feelings, portray "us" in a kind of allegory. The signature of the animal is somehow more apt than the colossal hieroglyphic of the rocks, the silent autograph of the plants, or the calligram of the landscape itself. Like amusing, wise, terrible, curved mirrors, animals prefigure human society. The lion, for example, shares a primordial ecology with humankind, a long history of symbolic power linked to our own feelings "in the blood". Likewise, the bird is spirit and the snake is the earth of our most elemental self, our mundane world, and our imagination.
>
> (Shepard 1997: 10)

Distinctive to Shepard's approach is the way he combines his thesis about the fundamental significance of animals for humans with an awareness that historical and cultural change impacts greatly on whether that significance is sustained or undermined – it being fundamental is no guarantee that it will be recognized as such and that the conditions necessary for animals to go on playing this role will prevail. Sometimes cultures change to the effect that the figures Shepard exemplifies in the passage above gain in complexity. Or change is such that they lose their immediacy, become distanced from their origins – say, in first-person encounters with particular animals, conveyed for generations to come in the stories told about such encounters – by the loss of tradition, a loss nowadays typically caused, and speeded up, by the widespread urbanization mentioned earlier. Despite these strong trends, Shepard is adamant that as the images and recollections of some animals become obsolete, others will emerge. "More than monuments to human imagination", he writes, "the whole panoply of their mythic, fantastic forms is based on thousand millennia of watching and studying real, wild animals. This creative perception of animals is in us still, a perennial satisfaction and pleasure, one of the oldest human vocations, building on the complexity of natural history, limitless not only in gathered facts but the feeling that one is part of a gathering, a new understanding, a deepened participation" (ibid.).

I read these statements as a warning against being misled by Shepard's claim about a deep affinity between poetry and animals. Our relationship to animals, he insists, is not first and foremost poetical or symbolic; not esoteric or aesthetic, not about abstraction and what feeds our powers of abstraction. He even goes so far as to dismiss nature writing – if anything, more popular today than when he made his comment in the 1990s – on the grounds that it nourishes a view of nature as "esthetic abstraction", breeding in the writer "the greed for literary reputation, captivating its readers with a spurious substitute for experience in the natural world" (1997: 11).

The fact is that there is no such substitute. That this is so is the unmistakable message of the greatest modern classic of them all, Rachel Carson's *Silent Spring*, one that Shepard fears is subject to another type of misreading, namely the managerial and technical one of getting rid of pesticides, of ensuring that they are replaced by some technological regime in agriculture that is less toxic for the animals and humans affected. This is no better than the seemingly obverse danger – or temptation – of aesthetization, turning the natural world into a beautiful painting on the wall, helping deceive its admirers into thinking that everything is well, that nature's beauty is some kind of timeless sublime object – indeed, art, made for art – of the sort Kant theorized in his third *Critique*. Only in an immediate, utterly concrete sense was Carson's book a warning against the effects of pesticides; more fundamentally, it was a warning against "the deafened self, against emptiness". Again, what this is about is resisting being distracted from the one, all-important issue – namely, what to make of our encounter with the animals, meaning what to make of it *today*, when "we ourselves prosper in unseemly numbers as they vanish", forgetful of the possibility that "in the end our prosperity may amount to nothing without them" (ibid.).

So it is not only that there can be, philosophically speaking, no context-free, *sub specie aeternitatis* inquiry into why animals matter. To put it more precisely, while there may indeed be one answer as to their mattering, the conditions for their actually doing so may now be changing so rapidly and to such an extent that *the continued mattering of animals can no longer be taken for granted*. In other words, what Shepard urges that we appreciate without distractions and surrogates is in a state of being spoiled, of being lost. That this is so is not primarily about numbers. The fact that we humans become ever more and the animals ever less is of course a crucial part of the picture, illustrating the trajectory we seem to be locked into. But the numbers, their empirical correctness notwithstanding, may themselves distract us, rendering what is all-too-real abstract, so many items in a statistic (e.g., the number of wild animals in the world is halved in the last 40 years) we have no way of connecting to.

To read about the ongoing extinction of species and loss of biodiversity is not to understand what these changes in the natural world amount to. Why? Is there a gulf between what we know and what we understand?

Quite the contrary if we look at it from a historical perspective. As long as the acquisition of knowledge – say, about particular animals – is directly dependent on experiencing first-hand the animals in question, the knowledge gained is not detached from its object. Encounters between knower and known provide the check on the validity of what is assumed to be true to the reality. That this is so is borne out exemplarily in hunting. Hunting is a practice that can only succeed if the assumptions of the predator about the prey at hand are correct. This holds true regardless of the species identity of the parties: animals who prey need to get it right just as much as humans do, and, again, this holds regardless of whether the prey is an animal or a human. In this setting, assumptions are typically about what response to expect from the prey, whereby the nature of that response needs to be anticipated, and thereupon adapted to, in as exact a manner as possible. And the basis for that sequence – anticipating the other's response, and adjusting one's own actions accordingly – is experiential, it is based on the knowledge gained about the behaviour of that sort of animal in that sort of situation on earlier (relevantly similar) occasions. Far from infallible and foolproof, the knowledge thus acquired through previous experience is the only basis for action to rely on. Getting the knowledge and the assumptions-cum-anticipations about likely response right is a matter of life and death – who knows, perhaps this is the origin of the saying "getting it dead wrong"?

I will highlight a couple of things about this notion of knowledge before returning to Shepard's take on it. First, the context of discovery and the context of application are co-extensive and equiprimordial – both are *in situ*, that is to say, a sort of knowledge only to be acquired in the context for which it is – uniquely – suited and in which it is to be employed, adhering to the logic of trial and error, at that. To know is to learn how to put that knowledge to use in the particular case. This learning is a practice – something practiced – and as such inseparable from the *in situ* acquisition of the skills required to act on the knowledge. If we stick to the case of hunting, the skills it takes to approach the animal, locate its exact movements, and anticipate its most likely next move, depending on mine and/or that of other animals present on the scene, are skills to do with movement, with the optimal timing of the next move, based on the skills it takes to identify the moves and the accompanying sounds: a branch being broken high up in that three – what can it be? Should I wait or should I proceed? When is the time to move slowly, and when to move quickly? How is that absence of any sound, that sudden silence to be interpreted – has the monkey fled so that there is nothing to be heard, or is it staying put, hoping I will be deceived?

In situ knowledge is knowledge concerning something particular, in some particular place, for some particular subjects and suited to some particular pragmatic aims, most basically: survival. The coexistence on earth of animals and humans is fundamentally wedded to the same reality principle as the coexistence of animals, that of eating and being eaten. For all the timelessness and universal validity of these options as given conditions of existence for humans and animals alike, the knowledge they necessitate is anything but universal in the sense of *episteme*; knowledge, that is, of a general and apodictic type, such that 3 + 2 will always yield 5 as its one valid (true) answer, regardless of the entities making up the content of the numbers, regardless of who asks the question, in what time and place and for what practical purposes. This is knowledge of the "each and every time" kind, abstract and generalizable in that its validity is eminently transposable from the one instance where 3 + 2 is asked to the next: abstract because context-independent in that strong sense, making its subjects as well as objects (entities) interchangeable. By contrast, the kind of knowledge at play in hunting is designated "local knowledge" by anthropologists, depending for its validity on a fit with its circumstances in the particular case. Knowledge that has its basis in experience is knowledge that allows for the individuality of all involved. Two hunters who have shared the same kind of experience with a particular animal may differ in what they make of that experience in a future situation. Importantly, this also holds for the animals involved, being subjects in their own right and thus misconstrued as "objects" devoid of agency (more on this below). Both humans and animals incorporate new experiences into past ones, making for biographies that are unique to the individual. Yes, we may respond to the same situation, the same thing; but we do so from different points of view and with different baggage.

I shall have more to say about this later. As indicated, Shepard's take on this topic is somewhat different from the ethnographically informed one I have just sketched (see Descola 2013; Kohn 2013; Viveiros de Castro 2015). Shepard's perspective is cosmological, highlighting the evolutionary origins *in animals* of the supposedly unique capacities of humans for abstraction, for engaging in symbolism and in thought in Hannah Arendt's sense of "making the present absent, and the absent present" (1978), thus transcending in thought and going beyond in imagination what is bodily and sensuously accessed as the givens of any concrete here-and-now. Whereas we are physically bodily placed so as to access by way of our senses one particular portion of the physical world in any given instant, we may be thinking about something else entirely, something that has nothing to do with what is actually present, everything to do with something that is absent, say, something that was but is no longer, or something that has never been but hopefully one day will be, etc. Thought is free in this sense of transcending the spatio-temporal circumstances of its subject-object constellation.

In speaking of this as something recognizable for us humans, I do not mean to pronounce on whether thinking as defined here is an exclusively human prerogative – that is another issue. What interests Shepard is *animal-human continuity*, not division, in keeping with his main idea that only animals make us into humans. To appreciate his approach, consider this passage:

> We were latecomers in a well-established process that had gone on for fifty million years. The four-legged carnivores and their prey had long since learned that an animal, watched long enough, gradually dissolved into signs. It left the marks that came to represent it: footprints, urine, secretions, feces, molted antlers, scratchings and rubbings, gnawed stems, bones, feathers, beds, diggings, nests, tracks, and bits of fur as well as an immense range of sounds and smells available to us. Language and art meant that those signs, as "objects", could be transported and therefore shared with those who, for example, had never come upon the bleached leg bone of an antelope near a water hole, broken by hyenas, pitched by vultures, and chiseled by mice, but who could hear the story and see its representations for what it was in the wilderness and what it might signify as an analogy of a seamless social form. Our human specialty was to dislodge such signs from a momentary stuckness in place and time and build a mental world of them that could be played over at

> times of our choosing. As signatures these animal signs were the beginning of abstraction and symbolizing in a drama we primates already "knew" was social at heart: gestures, expressions, innuendos. In less than three million years, all these categories of the self and society were shaped by the traits of animals observed, the dangerous, competitive, beautiful, tasty, scrounging Others.
>
> (1997: 24)

Shepard's foregrounding of the role played by animals before humans arrived on the scene, as well as their persistent role in providing the foundation from which distinctive human capabilities and practices spring – i.e., the function of language (first oral, than written) in socialization and integration – is conspicuously absent in the work of, say, late Habermas. If animals are at all mentioned, it is to argue a point about what they cannot do as compared with what humans can do. Animals serve to provide a sharp relief against which assertions about human exceptionalism can be made. In discussing the research of Michael Tomasello, Habermas speaks of the equiprimordiality of "the three human monopolies": the employment of symbols, mutual adaptation of perspectives and intentional attitude to objects, and seeks to show how they are drawn upon in the core phenomenon of communication, defined as the ability of humans to come to agreement with each other about something in a shared epistemic world, drawing on the resources available to them from a shared symbolic lifeworld. Animals are mentioned but once in the course of making this argument: "Chimpanzees lack the kind of communication where a common objective world is accessed by virtue of the meaning-identical use of symbols within the horizon of an intersubjectively shared lifeworld" (2012: 65; see 2019: 241).

Let us grant that the use of symbols is unique to humans. It is so because it involves convention, so that a symbol refers to its object not directly, as do icons, but indirectly, by virtue of the ways in which they relate systematically to other symbols, where their doing so is a matter of agreement among those who use them, and a genuinely social and cultural attainment at that. What many anthropologists question, however, is the widespread assumption that all representational phenomena have symbolic properties, an assumption that amounts to *a conflation of representation with language*, whereby language is taken as per definition involving symbols and so as involving capacities and facilitating practices presumably unique to humans. What we risk ignoring here, as Eduardo Kohn seeks to show in his ethnographic study *How Forests Think*, is that non-human beings and the lifeforms they constitute also *represent the world* (2013: 8, 40). Animals like the monkey mentioned above orient themselves by way of signs, be it by making them or responding to them in their environment. Indeed, the monkey relates to what Peirce called indices, which, unlike the icon that is a likeness to the object it represents, point to something else – be it an absence, a significant one at that. So the monkey is able to make connections between some event (a sound made by another animal fleeing upon hearing the sounds made by hunters approaching) and another potential one that has not yet occurred – the fleeing animal is interpreted as a sign to the effect that something – potentially dangerous – is about to happen; to prevent what is a mere possibility from becoming a reality, response in the form of the monkey's decision to flee itself is called for, and hesitation to do so means death.

This may appear to be an aside, but it is not. Even if we grant that symbols are the one kind of sign – the two others being icons and indices (Peirce) – found only in humans, the chimpanzees Habermas invokes by way of contrast do possess the capacities to direct their attention at something shared, and interpreted in like manner, in a likewise shared objective world (environment) and by drawing on the resources of a shared lifeform. To learn what something signifies so as to be able to interpret that thing (object, event) as meaning the same to me and to others, eliciting the same sort of response, is a sign-employing process that is just as vertical *and* horizontal in monkeys and chimpanzees as it is in humans, to use Habermas' own terms: these animals no less than us humans are simultaneously engaged in relating to something shared out there (vertically) and – meaning: as – something shared among themselves (horizontally), in each instance both presupposing and confirming a sharedness of interpretation and response that is a *social* achievement, one that each individual (animal, human) can be part of as a member of a society of interpreters that by way of example sets the standard for correct interpretation: "these animals are prey to us, those are predators"; "that kind of sound, made by that kind of animal, means danger", etc. Gestures play a crucial role in stabilizing this entire practice, and in the socialization into becoming a skilled participant, by conveying what expressions and what movements mean what, and so help direct apt response. This is what the infant learns, and can only learn, from its significant others. In that sense, it is a social phenomenon, the appreciation of which can be traced back to Darwin's classic study *The Expression of Emotions in Man and in Animals* from 1872.

Against this background, we may turn to the question indicated above: who may qualify as significant others? Who is indispensable? Members of the same species only, or members of other species as well? If it takes a village to raise a child, does that village only consist of humans?

VI

I propose at two-step approach to these questions. We begin by looking at the matter from a historical perspective, then move to an ontogenetic one.

We saw above that Shepard holds category making and the emergence of taxonomies to be based on animals, tracing the evolution of the human mind and the beginning of language to an origin shaped by the observations of and encounters with animals. The life-or-death deciding necessity of identifying all sorts of creatures based on the sounds and moves they make and the traces their presence leaves is common to animals and humans, and historically it is a phenomenon that long antedates the existence of humans on earth. In developing what will become the distinctly human way of engaging in these practices, then, humans don't start from scratch; human societies start by learning the basics of from so many non-human creatures and life-forms, adding, as we saw, the employment of symbols, the key element in human language qua oral as well as written.

Shepard's take on this topic differs from Habermas' in being utterly concrete. He gives pride of place to the question: what sort of entities are being identified, are being counted, inasmuch as they are the entities encountered in the world the human agents in question have in common? How comprehensive and complex a world are we talking about here? How rich or poor in terms of diversity along different axes, species diversity being one, biological diversity a second, cultural diversity a third? And how to answer such questions? Yes, by comparison; but diachronic or synchronic?

To be sure, what I am getting at is so much better expressed concretely than abstractly. "Tribal peoples around the world", writes Shepard, "know hundreds of plants and animals by name and natural history. The Nuba of Africa identify more than forty species of locus (biologists recognize only ten) and twenty-seven varieties of sorghum which are botanically but three" (1997: 54; 1998: 202). He quotes an anthropologist who observes that "in the two preliterate societies in which I have carried out field research, knowledge of the biological world constitutes – I would claim – a greater chunk than all other types of knowledge combined" (ibid.). He calculates, Shepard tells us, that primitive tribes have an average of 1,000–1,200 kinds of plants and animals in their vocabulary, and he makes the point that familiarity with this great diversity of organisms is not primarily because of their economic usefulness. Whence does it spring, then? Shepard suggests that such knowledge is "good for its own sake", owing to its inherent merits. The single universal characteristic here, says Shepard, is that the organisms are named, and this naming – perfectly without God's doing it – makes it possible to "handle" them cognitively, that is to say (and approaching now, but only now, Habermasian territory), making them objects of discourse.

But that is not what comes first or is most important. Naming is a delight, even an end in its own right; if not for animals, then certainly for humans. Shepard mentions Merrill Moore, a lifelong shell collector who, exploring his taxonomic frenzy beginning in his childhood, asks himself: "Why do I like shells?" Bringing home the shells he found on outings, handling and examining them, seeing his collection grow – all of this provided visual and tactile delight, "reminding him of places and outings, and opened up the complex lives of the mollusks themselves who seemed to him separate personalities, 'like people'; shells are 'a world within a world'; if I wish to see richness in Nature, variety in pattern, and seemingly endless alterations in colour and line, all I need to do is take them out and look at them" (1997: 56f.). Shepard comments that collecting found objects "disengages things from their places and their context"; it deprives them of their *aura*, to use Walter Benjamin's term, and yet it bestows on them the status of things to be valued in their own right, to be admired and enjoyed aesthetically (ibid.).

That said, animals are more central to the fundamentals of identifying by naming than are shells. For us, Shepard maintains, animals are distinct but related entities. They are alive like us, yet they exhibit that liveness in ways so different from us, and – no less importantly – from each other. In the practice of naming, the first criteria are anatomical; instinctively we identify animals according to general body form ("duck"), only to specify further the identity by specific cues ("green-winged teal"). Acquisition of this knowledge – first general, then more specific – is an eminently concrete undertaking, having physical encounters with the animals in question as its context of discovery and verification – here, nothing can replace the primary access provided by the senses, meaning the full repertoire of them all: seeing, hearing, smelling, tasting, touching the animal, getting to grips with its species-specific traits by way of observation of its behaviour in its natural habitat and with its natural others.

Since time immemorial, animals have played a prominent part in children's games, leading Shepard to ask: Why is there such an apparently universal need for us humans to see different aspects of ourselves in animals instead of in other people? Or more to the point, why is the attention to aspects of ourselves as exhibited in animals so important to our coming to understand these aspects in ourselves and in other people? Why aren't animals instead a detour in this respect?

"Each animal in play", Shepard writes, "reveals a certain trait or feeling exhibited in its behavior. Each kind of animal gives concrete representation to an ephemeral and intangible element of the human self such as assertion, intimidation, affection, doubt, determination, kindness, anger, hope, irritation, yearning, wisdom, cunning, anticipation, fear, and initiative. Only when these feelings [and states of mind, AJV] are discovered outside the self and then performed can such intense but elusive 'things' be made one's own" (1997: 83). In declaring to be a fox, the child knows very well that it is merely playing that animal; doing so is not to rehearse an actual role in its social life but to learn the more fundamental lesson: that there are roles. In this process, the child is in no illusion about being the fox; the difference between animals and humans is maintained not glossed over. Still, animals – and some more than others, depending on the role in question (foxes as cunning) – are perceived as similar enough to humans to have this function. Shepard puts the point like this: "The miming of animals in play does not animalize the child, since it does not teach us to live in holes, run on all fours, or catch mice in our mouths. Nor does it humanize the fox, as even the child knows in its heart that the whole matter is 'just a story'. It is important as a joint adventure by children, part of the biology and culture of childhood, in "capturing the Other" in order to constitute a self" (1997: 84).

Shepard corroborates his argument by Konrad Lorenz's observation that all mammals learn by rote memorizing, whereby imitation – according to Aristotle the basis from which all learning, all acquisition of knowledge as well as skills springs – is absolutely instrumental. As we saw when referring to Eduardo Kohn's ethnographic work, there are three stages or "modes of representation": enactive, iconic, and symbolic. Children are committed to the first stage, acting out what they feel, so that to feel something is tantamount to enacting it. Thus children learn to walk like mice learning to run a maze. They learn that an object hidden continues to exist even though they cannot see it. As for manual skills, they are not taught but learned by watching and imitating. Again, there is no substitute for full-fledged bodily and sensuous engagement with the world, accessing it by way of the primary senses, the reliance on which was strictly forbidden by Descartes, who thereby closed the principal door of humans' access to other creatures. The practice of watching and imitating is found in other large mammals too, but the human repertoire, Shepard writes, is peculiar in that "children imitate not only other people but other species, a characteristic of our own kind of consciousness" (1997: 87). Again, being an animal and acting the animal are not the same, and the child knows the difference. What we need to recognize is that such enactments are "transformative traits of our humanity". And we do not, upon becoming adults, leave them behind; "the skills of categorical thought by means of the taxonomy of other species is never completely put by; we continue as adults to extend our knowledge of the nomenclature of plants and animals" (ibid.).

At this juncture, Shepard addresses the notion of anthropomorphism, understood as attributing human qualities to non-humans by way of projection. Like Midgley, Shepard is sharply critical of the critique of anthropomorphism where it is sought eliminated as a sort of "category mistake" that, while evidently springing from a deep-seated disposition in us qua species, in fact prevents us from any objective and scientific understanding of animals. The mistake made by its critics old and new, argues Shepard, is to suppose that anthropomorphism is an end in itself; "they judge it without regard to its function in childhood as a precursor to poetry and other metaphor". He goes on to quote from psychoanalyst Bruno Bettelheim's classic *The Uses of Enchantment*: "If we do not understand what rocks and trees and animals have to tell us, the reason is that we are not sufficiently attuned to them. A child is convinced that the animal understands and feels with him, even though it does not show it openly" (1997: 88; see Bettelheim 1976: 46). Properly understood, then, anthropomorphism is about gaining a *first access*, a full-fledged sensuous entry into the feelings, states, and responses – the inner life – of an animal, taking these features to be sufficiently similar to those the child finds in him – or herself to enable a tentative appreciation of what is supposedly "going on" in that animal in that situation, often by way of its reacting directly to the child's action towards it, highlighting the eminently physical encounter that provides the setting here. Only by acting upon this assumption of sufficient *likeness*, an assumption unquestioned during childhood, is the child in a position to develop, gradually, an appreciation of the *dissimilarities* existing between itself and that animal. The projection involved in anthropomorphism is not, as Shepard rightly stresses, the end point but the indispensable point of departure.

What anthropomorphism helps accomplish, then, is something truly precious: the recognition of continuity between animals' lives and our own; a continuity that "reinforces a profound and enduring metonomy, a lifelong shield against alienation" (1997: 88). At the end of puberty, which marks the end of innocence, we begin "a lifelong work of differentiating ourselves" from the animals, a work that could not commence, as we just saw, were it not for the way it grows from an earlier, unbreakable foundation of contiguity" (ibid.). Barring that, to insist on ourselves simply as different from the animals would deny "the shared underpinnings and destroy a deeper sense of cohesion that sustains our sanity and keeps our world from disintegrating". Herein lies the fundamental importance of anthropomorphism: it "binds our continuity with the rest of the natural world" (ibid.). Accordingly, "failure to nurture childhood enthusiasm for animals produces adults bereft of diverse living forms as the metaphorical basis of religious conceptions and values. Such a society fails to honour the underlying unity of diversity because it lacks experience of it" (1997: 89). Fundamentally, and despite the modern talk of "humans or animals", we are "not part of a binary duo or of a simple juxtaposition" with animals. "However we may define ourselves as a species", Shepard maintains, "the final act is to recover our animalhood, to see all taxonomy for what it is: a means, not an end, to thought" (ibid.).

VII

Above I asked who should qualify as "significant others" in the development of human individuals.

So far, we have looked at how Shepard approaches this question from a general perspective, acknowledging the evolutionary role of animals in providing the basis for the specifically human capacities in such practices as identifying entities by naming them. Animals, whose existence on planet earth long predates that of humans, are the prototype of such entities, holding as we noted an instinctive fascination in simply *being alive like us* and thus inviting comparison that is equally open to difference and similarity.

However, ever since their beginnings in the nineteenth century, academic psychology and sociology Western style have to a large extent ignored the significance of the human-nature relationship, leading commentators to speak of the "nature phobia" characteristic of modern social science (see Benton 2001). I shall have more to say about this bias in later chapters.

In turning now to Shepard's account of ontogeny as the dimension where the relationship between the human individual and the wider natural world, including animals, is played out, a statement Shepard quotes from psychoanalyst Harold Searles is pertinent: "It seems to me that the highest order of maturity is essential to the achievement of relatedness with that which is *most unlike* oneself" (Shepard 1982: 14; see 1998: 213). The quote is from Searles' study *The Nonhuman Environment in Normal Development and in Schizophrenia*, published to little acclaim in 1960 and only now beginning to be recognized as a classic. The "world" that matters to human development from its very beginnings is a non-human one as well as a human one, pointing to the "total environment" within which socialization and identity formation take place. "The human being", writes Searles, "is engaged throughout his life span in an increasing struggle to differentiate himself increasingly fully, not only from his human but also from his non-human environment, whole developing in proportion as he succeeds in this differentiation an increasingly meaningful relatedness to both environment and people" (Searles quoted in Shepard 1998: 183). The critical thesis that the atypical psychiatrist Searles shares with Shepard is that the non-human dimension of that total environment has become so marginalized in contemporary Western society as to render fellow humans the only creatures vital to human development and the relationships it consists of. In that sense, to become human has everything to do with culture, that is, with other humans and with the artefacts and products of humans, and nothing to do with the beings making up the wider natural world.

Many will regard this thesis and the critique it offers as exaggerated and short of a substantive case. So let us consider the argument, sketched very generally so far, in more detail. In doing so, we shall turn to the book for which Shepard is probably most well known, *Nature and Madness*, aptly titled so to signal his affinity to Searles' overriding project: to demonstrate the link between human sanity and the natural world.

Shepard starts with an evocative depiction of the original scene:

> For the infant as person-to-be, the shape of all otherness grows out of the maternal relationship. Yet, the setting of that relationship was, in the evolution of humankind, a surround of living plants, rich in texture, smell, and motion. The unfiltered, unpolluted air, the flicker of wild birds, real sunshine and rain, mud to be tasted and tree bark to grasp, the sounds of wind and water, the calls of animals and insects as well as human voices – all these are not vague and pleasant amenities for the infant, but the stuff out of which its second grounding, even while in its mother's arms, has begun. The outdoors is also in some sense another inside, a kind of enlivenment of that fetal landscape which is not so constant as once supposed. The surroundings are also that-which-will-be-swallowed, internalized, incorporated as the self.
>
> (1982: 7)

It is hard to read this passage without a feeling of nostalgia for a "natural" setting now gone, one not to be taken for granted in any way, and hardly anywhere, in today's world. For many a reader, the label "nostalgia" will be intended as a criticism, akin to "romanticizing" bygone times and societies, from which we moderns have little or nothing to learn. But if, as suggested, the topic at hand is to do with the preconditions for sanity, and thus for psychopathologies and their specific origin in the way we relate to the natural world, to the total environment of which we are part and by which we are affected, unconsciously as well as consciously, then dismissing Shepard's approach is not only premature but risks robbing us of important insights. What could they be?

Historically speaking, the transition from a setting pervaded by the abundance of eminently proximate, literally hands-on, seeable, hearable, smellable, touchable non-human creatures of innumerable sorts to the present-day one of the paradigmatic urban setting is a story in many steps. The transition from hunters, and hunter-gatherers, to societies organized around agriculture, from the nomadic to the stationary, from the land to the cities, from low-tech to high tech, from a situation where human groups were few and far between amidst the endlessness and vastness of seemingly all-powerful, omnipresent nature (wildness) to a situation where this proportionality is being rapidly, and globally, reversed – this multilayered transition can only be invoked as a well-documented historical fact here, not accounted for in detail.

Concentrating on the ontogenetic dimension, Shepard asks what difference it will make if the first nine years of a child's life are played out in a fabricated environment as distinct from the nature (outdoors) setting depicted above. This is what he has to say:

> The absence of numerous nonhuman lives, a variegated plant-studded soil, the nearness of storms, wind, the odors of plants, the fantastic variety of insect forms, the surprise of springs, the mystery of life hidden in water, and the round of seasons and migration – how does the lack of these really hurt the child?
>
> First, life in a made world slowly builds in the child the feeling that *nonlivingness is the normal state of things*. Existence is shaped from the outside or put together. Eventually he will conclude that there is no intrinsic unfolding, no unique, inner life at all, only substance that, being manipulated, gives the illusion of spontaneity.
>
> Second, when he goes back into society after adolescence, having missed the initiation into the world of final mysteries and its poetry, he will believe either that (a) all life, including people, is, in fact, machines or (b) the only truly living things are people; this opens the door for all the dichotomies that separate the human from the nonhuman on the basis of soul, spirit, mind, history, culture, speech, rights, right to existence, and so on.
>
> Third, he becomes either a spectator or an exploiter. The world may be interesting or dull and useless, but it is not one that feels or thinks or

> communicates, that has special messages for him, or that has independent purposes of its own. The world's inability to feel is his own mechanical response projected upon it. He is like the infant whose emotions are crippled because he was suspended too long in an inert world.
>
> (1982: 103; my italics)

In giving this critical account, Shepard is well aware that the explanatory model he relies on – ontogeny repeats phylogeny, so that each child repeats in its development the evolutionary history of the human species – is dismissed as speculative by many scientists today (see 1997: 6). Whether seen as depending on that model or not, the idea of a sick society that Shepard implies here is one that allowed Freud, in his late work *Civilization and its Discontents* (1930), to contend that, "under the influence of cultural urges, some civilizations – or some epochs of civilization – possibly the whole of mankind – have become neurotic" (Freud quoted in Shepard 1982: 4). Erich Fromm would later take this proposition as the point of departure for his widely read book *The Sane Society* (1955).

More relevant for my purposes is Herbert Marcuse's *Eros and Civilization*, published in 1956, a study that covers much of the same territory as the more famous *Dialectic of Enlightenment*, authored by Marcuse's colleagues within the Frankfurt School, Max Horkheimer and Theodor W. Adorno. Relying on the contributions of late Freud in a more rigorous manner than his colleagues, Marcuse explores the paradox that came to define decades of work within Critical Theory: that the enormous progress made thanks to the scientific rationality of Western civilization was being paid for by a series of psychopathologies, revealing the celebrated progress as a prison as far as the realization of humanity's potential for happiness and freedom is concerned. Under the imperative of taking part in the rational transformation of the human and natural environment, the ego was itself transformed, namely into "an essentially aggressive, offensive subject, whose thoughts and actions were designed for mastering objects" (Marcuse 1956: 109). To be a human subject was to act as such *against* a world of objects. "Nature", Marcuse contends, "its own as well as the external world, were "given" to the ego as something that had to be fought, conquered, and even violated – such was the precondition for self-preservation and self-development" (1956: 110). The individual is involved in a two-front struggle: one internal (intrapsychic), aiming at the subjugation of its sensuous and appetitive faculties, condemned since Plato as the "lower" ones in being also found in animals, and since Descartes owing to their alleged role in leading us astray epistemically, in our effort to access external reality in a reliable, objective, and detached manner, as required in philosophy and in science alike. The other struggle is the external one, where nature is *a priori* experienced by an ego "bent on domination and therefore experienced as susceptible to mastery and control" (ibid.), a stance that Habermas (as we will see more closely in Chapter 3) would come to designate as constituting a "quasi-transcendental cognitive interest", the one "rational" attitude for humans as a species towards the so-called objective world, regardless of its differentiations between animate and inanimate, organic and inorganic, including animals (see Habermas 1972).

In Marcuse's terminology, what Freud helps us understand is that in Western modernity the "performance principle" adopted and internalized by the ego by way of its two-front struggle is made into the reality principle as such. But Freud failed to realize that, historically speaking, the oppressive "modification" of the pleasure principle – that of eros, of uninhibited sexuality and play – by the reality principle in its present form has in fact become "unnatural" and unnecessary: "Freud's consistent denial of the possibility of an essential liberation of the pleasure principle", argues Marcuse, "implies the assumption that scarcity is as permanent as domination – an assumption that seems to beg the question". Contra Freud, then, "the historical possibility of a gradual decontrolling of the instinctual development must be taken seriously, perhaps even the historical necessity – if civilization is to progress to a higher stage of freedom" (1956: 134).

Despite his criticism, Marcuse credits Freud's concept of "primary narcissism" with the important idea of "a non-repressive erotic attitude toward reality", taken as "an undifferentiated, unified libido prior to the division into ego and external objects". Primary narcissism, Marcuse maintains, is more than autoeroticism; "it engulfs the 'environment', integrating the narcissistic ego with the objective world". He quotes Freud to the effect that "the ego-feeling we are aware of now is only a shrunken vestige of a far more extensive feeling – a feeling which embraced the universe and expressed an inseparable connection of the ego with the external world", i.e., what Freud termed "oceanic feeling" (Marcuse 1956: 168; Freud 1930: 13ff.). Far from being an egotistic withdrawal from reality, proof of immaturity or a neurotic symptom, this type of narcissism "denotes a fundamental relatedness to reality which may generate a comprehensive existential order" (1956: 169). This, and nothing less, is the revolutionary potential Marcuse spots in Freud's concept.

There is no denying that, for present-day readers, there is something dated about Marcuse's attempt at a synthesis between Marx and Freud, with his insistence on making a distinction between necessary repression and "surplus repression", between repressive and free sublimation, and on the "possibility of libidinous work relations", taken as helping attain a "non-repressive society" where gratification is finally liberated from its constraints in the psychic prison created by the imperatives of the hegemonic reality principle, obsessed with mastery and control in all areas of existence.

Even though Shepard does not mention Marcuse's work, he helps us see why it must be considered a missed opportunity. Let me put it like this. What was left largely unexplored in the work of Freud and Fromm is so also in Marcuse: a systematic exploration of the way in which the socialization of the human individual is related to its experience with the non-human. Only with the publication of Searles' study in 1960 would that become a topic in its own right. It is symptomatic of the limitation I have in mind that not a single passage in Marcuse's study is devoted to the role played by animals in making us human and thus the importance of their decline in accounting for how "a sick society produces sick individuals". While no reader of Marcuse will be in doubt about his fascination with the natural world as a site of uninhibited human sexuality, non-material (spiritual) gratification and a sense of profound joy in being alive, "nature" remains a general and abstract concept, never broken down into the concreteness of human-animal encounters that informs every page of Shepard's books. We are urged to understand that the attitude of mastery and control is deeply wrong and deeply stifling to the humans caught up in it, instrumentalizing themselves in the process of instrumentalizing nature. But the wrongness is never considered from another point of view than the human-centered one: the question about its stifling consequences for the other-than-humans affected is never raised.

According to Shepard, the child whose socialization takes place in a human-made, urban environment that allows minimal exposure to other than human forms of life is, in effect, asked to proceed directly from its symbiosis with mother to a mastery of social relations. This means that the child skips the genetic interlude in this developmental task, in which it indulges for eight to ten years in nature, going directly as it were to the real job of life. From the age of four to the age of ten, things found in nature will be picked up and played with by the child – for example, the broken piece of a branch on the ground will immediately be turned into a sword. Although it takes fantasy – imagination – to turn that natural entity into an artifice (a weapon) to play with, the idea that *things encountered in nature are ideal for this function* is instinctive in the child. The natural world is taken as fully fit for this sort of spontaneity, as – literally – the paramount natural setting for it. The child's acted-upon assumption is that nature is not alien to its human needs and capacities – be they cognitive, be they emotional – but conducive, even indispensable, to their formation and enactment. Such immersion in otherness – distinct from alienness – has no substitute in the array of images and toys that so many children today are exposed to in the form of the "pictures, zoos and gardens, decorations, Disney films, motifs and designs" that Shepard describes as "a stew of nature so arbitrarily presented that the results of the child's years of trying to fix it in his heart will only lead to despair" (1982: 105). In this case, there is no other sword than the one the child receives from the parent who bought it in a shop. It is being made of plastic instead of wood, its being something received as ready-to-play-with and with the tag "sword" on it, cheats the child of the chance to create and to invent, of the experience of finding the item-to-be-made-into-sword by way of the pretence, the *as if*, that is the essence of true play. No wonder, remarks Shepard, that the child of 13 years who has been cheated of this experience and what it actively requires of him, turns with such keen interest to machines (ibid.). (Note that Shepard clearly has boys rather than girls in mind here, without commenting on it.)

Shepard says that he "suspects" that "the substitution of animal toys and pictures as a kind of appeasement damages the individual's *social* relations, that it convinces him at some deep level that the given state of man is without pattern or purpose, that men have unlimited capacity to fabricate things" (1982: 105), and that, we may add, it has always been like that, meaning that a child who grows up in an environment that is completely (or close to completely) human-made will have no idea as to how recent and indeed unprecedented this situation is. The tendency to take for granted that relationships are such as to consist of fellow humans only and that to the extent that relating to non-human entities does at all figure, they will be artefacts not animals, made not grown, is exemplified in Habermas' comment about a two-year old child who, in its interaction with an adult, finds it amusing to "use a pencil as if it were a tooth-brush"; the role played by animals, as opposed to artefacts, is analytically as well as philosophically ignored (see Habermas 2012: 64). This *silencing in theory* of non-human others of which animals are an example both follows from and in turn helps reinforce their *silencing in practice*, in empirical reality, rendering the theory in question uncritical in that it merely mirrors, and portrays as the way things simply are, what in reality is the result of enormously important historical and cultural change. This is my way of putting it, not Shepard's. But I think it expresses the crux of his project.

Shepard's admission that his thesis will come across as speculative does not prevent him from putting it in strong words:

> Urban civilization creates the illusion of a shortcut to individual maturity by attempting to omit the eight to ten years of immersion in nonhuman nature. Maturity so achieved is spurious because the individual, though he may become precociously articulate and sensitive to subtle human interplay, is without a grounding in the given structure that is nature. His grief and sense of loss seems to him to be a personality problem, so that, caught in a double bind, he will be encouraged to talk out his sense of inadequacy as though it were an interpersonal or ideological matter. Indeed, the real brittleness of modern social relationships has its roots in that vacuum where a beautiful and awesome otherness should have been encountered. The multifold otherness-with-similarities of nonhuman nature is a training ground for that delicate equilibrium between the play of likeness and difference in all social intercourse and for affirmation instead of fear of the ambiguities and liveness of the self.
> (1982: 108)

In viewing the increasing injury to the planet as a symptom of human psychopathology, and by implication as a process that will not come to a halt before the root causes of that pathology are identified and dealt with,

Shepard points to the self-reinforcing logic at work. If losing the connections to nature and its multitudes of non-human others leads to a situation where only fellow humans qualify as significant others and where artefacts replace animals, and pets replace wild ones, the indoors replace the outdoors, so that *reality in its entirety is domesticized*, a world of humans for humans, then there is also a loss of experiences such as to be able to *register* that lost connection to nature. Can we miss what we have never experienced? What about loss of a kind that antedates our own existence? How to assess the impact of absence? I shall come back to these questions in later chapters. Suffice it to say here that the resulting symptom is afflictions such as alienation, lack of meaning and of a sense of belonging; that is to say, the catalogue of "social pathologies" theorized by the pioneers of social science, by it Hegel and Marx, Weber and Durkheim, and not forgetting Freud.

On Shepard's account, these pathologies cannot be healed as long as their roots as well as remedies are sought in that one dimension now taken to define and exhaust the world as relevant to humans: the world consisting of humans where everything rotates around them. No matter how many people we are, and how spectacularly we succeed in putting our stamp on everything that exists, determining its meaning and worth or lack of such – our problem will not be solved, our pain not removed, insofar as we overburden the world qua human, qua social and cultural, with the task of fulfilling all our needs and enabling the flourishing of all our capacities qua human. The more one-sidedly obsessively we continue to bark up that one tree, the human-grown and human-made one, the more we neglect and, in the process, ruin that other-than-human dimension which, possibly, would provide what we seek.

VIII

Let us look at how this historical change affects human-animals relationships more specifically.

As we would expect, Shepard is highly critical of pets and the role that they have come to play in an era of dwindling populations of wild animals. In *The Others*, he quotes Walker Perry's statement that "the autonomous self feels so alone in the cosmos that it will go to any length to talk to chimpanzees, dolphins, and humpback whales. Why do people in general want to believe that chimps and dolphins and whales can speak…to establish communication with extraterrestrial intelligence?" (1997: 140).

Of course, the animals mentioned here are not pets – they are wild animals. Or at least they used to be, and to be treated accordingly. The point Shepard wants to make is that they are no longer so: there is a tendency to treat more and more animals along the wild-domestic-pet spectre as were none of them really wild, *as were they pets*. In light of the claims about the cultural taboo against anthropomorphism made above, this may seem surprising. And indeed, we need to explore the hypothesis that this taboo, in the strict version discussed above, is being lifted in contemporary society. In exploring this we will need to distinguish carefully between whatever scientific arguments there are for lifting the taboo, on the one hand, and the sort of emotional (sentimental) needs that underpin lifting it, on the other.

In discussing pets, Shepard's backdrop is our crowded yet increasingly lonely planet, one where the loneliness that existentialist philosophers like Sartre had posited as a metaphysical given, as defining the human condition in a timeless manner, has turned into a *literal* loneliness, namely that of humans in times of loss of all sorts of non-human others, amounting to a sort of loneliness that is historically unprecedented and as such as yet scarcely contemplated by philosophers. On my reading, the dialectic Shepard considers triggered by the factual and utterly concrete decline of animals in the wild is one of psychological compensation: the fewer animals left, the more important it becomes for us to make some contact with them, not only in the sense of acknowledging them as coexisting with us on Earth, but in the more demanding sense of *having the animals acknowledge us*, giving us some sign that they see us and that we are important to them, meaning that there is – or could be – some reciprocity to this need, that it will play out as a meeting-halfway between different species, us and them.

Not only is this a tall order; it is fundamentally mistaken. The point is not that humans are crucial to making animals into animals: we are not. Rather, and this makes for asymmetry and non-reciprocity, animals are crucial to making us human, and they are so in virtue of being different from us, not us and not like us, insists Shepard. Their being different means that they are indifferent to us; the more indifferent, the better, if I understand Shepard correctly. This is the whole philosophical point about animals being *sui generis* wild in Shepard's perspective. However, as we shall see in greater detail in later chapters, only animals that are in a position to maintain their independence from humans can be indifferent to us the way Shepard affirms. Should their difference-making-for-their-indifference be lost as their major characteristic, a *double loss* would be involved: the animals would lose their intrinsic identity and their intrinsic worth, tied as it is to what they are in themselves and not to what they are to humans; and in doing so, they would lose their significance for humans – the other-than-anthropocentric loss would be an anthropocentric loss as well. Note that the first loss is a loss in its own right and not a loss only to the extent that, and on condition that, it amounts to the second, human-related loss.

This is my way of putting it, overly theoretical, to be sure, and in need of making distinctions between different sorts of animals so as to do justice to the different ways in which animals are important in our becoming fully human in Shepard's partly cosmological, partly psychoanalytic sense, underscoring his work's originality. Shepard recalls that until the twentieth century, rabbits, chickens, ducks, and geese were still kept in backyards, even in the cities; "local fairs and markets had large livestock sections, draft animals were still abundant, farmers drove pigs and cattle to market down the streets, and knackers butchered them in alleys; dogs and cats ran freely in the streets" (1997: 143). This is an important observation that illuminates not only the abundance of animals, in numbers and types, that were part of the everyday lifeworld of, say, children growing up in cities as well as in a rural environment. It also illuminates how today's mostly indoor pets, like dogs and cats, lived more autonomous and less domesticized lives until factors such as increasing car traffic forbade it. In the course of the historical shift Shepard talks about, "domestic animals have gradually become surrogate companions, siblings, lovers, victims, workers, parents, competitors, oracles, enemies, kinfolk, caretaker-watchmen, and so on" (1997: 142), making for a self-reinforcing displacement within the crude category "animals" so that the wild ones more and more disappear from view and the domesticated become ever more so as to become pets. As we see, the sequence is essentially about the ever-growing loss of animals qua wild and other-than-human.

Shepard depicts the sequence as a drama in five stages. Act I is an outer circle of wild animal life which was a major focus of human attention; encounters with such animals as bears and wolves would inspire stories to be told again and again, with time assuming the status of myths, often tied to particularly charismatic animals, some of which came to be considered sacred. As such, Shepard remarks, the animals were seen as "conscious, unique, and different in spiritual power" (1997: 143). In Act II, certain animals were taken into captivity and, following the uses to which humans put them, their biological natures would over time conform to human dominance; taken in as co-members of the human household, they would become the domestic animals, such as horses and dogs.

For my purposes, Act III is the most interesting one. It begins, says Shepard, not with animals but with a class of things called "transitional objects". These – and an example would be teddy bear – are toted around by "anxious three-year-old children who are having difficulty becoming independent from their mothers and who are comforted by a soft object that is subjectively intermediate between themselves and the outside world" (ibid.). However, not all children require the security of such objects, observes Shepard; the children who do not are "those who are surrounded by abundant other forms of life" – that is to say, what used to be the normal situation in pre-modern society but is so no longer. The exact reason for the difference pointed to is not entirely clear. But "apparently animals in their diversity model a world of likeness and difference which makes the child's impending separation less frightening and also resonates with internal, psychic structures which can best be described metaphorically as a fauna. The stuffed toys [e.g., the teddy bear] are simultaneously huggable, transitional objects and 'animals'", helping to compensate children "for their lonesome social and ecological situations and preparing them for lifelong pet keeping" (1997: 144).

Act IV is the transfer of this affection from effigies to dogs and cats. And as the toys had been pets, the pets became toys. Act V completes the sequence in extending the equivalence of the living domestic pet and the stuffed wild toy to living nature: "If the domestic forms have all along been substitutes for the wild and the latter have become unavailable and unknown, it is easy to fuse the domestic and the wild", resulting in a situation where "the wild are simply those potential pets who do not happen to live with us" (ibid.). This is the background for John Berger's famous essay on why zoos disappoint. To be sure, humans get a chance to look at animals in the zoo. But the "look of animals" has vanished, tragically so even among the animals who were the wildest of them all – think of lions and tigers – in that they would epitomize the *being-different-from-humans and so being-indifferent-to-humans* that we saw Shepard consider to be their essential characteristic. The difference that makes this indifference possible and meaningful, even profound and challenging, is gone, dependent as it is on the place within the ecologies that define and energize these animals, leaving only an indifference that makes no difference to anyone. Animals thus made to be, to exist, out of place are no more lively than matter out of place.

Why do I find Act III particularly interesting? To begin with, "transitional object" is a technical term made famous in psychoanalyst Donald W. Winnicott's classic paper "Transitional Objects and Transitional Phenomena" (1951). Curiously, this fact is not mentioned by Shepard; despite his obvious debt to psychoanalytic theory, Winnicott is not mentioned in his book.

The intriguing part of Shepard's use of Winnicott's concept is to do with his (Shepard's) suggestion that the need for a transitional object in the form of a teddy bear, offering the three-year-old child comfort by being a "soft object that is subjectively intermediate between themselves and the outside world", is not a need felt by all children – there are those who, as we saw, can do without the security brought by such objects, these being children who are "surrounded by abundant other forms of life" (1997: 143).

The distinction among children made by Shepard, I suggest, challenges Winnicott's own use of the notion of "transitional objects" in the form of a pet like the teddy bear. Whereas Winnicott considers the need for the latter to be a "universal" one, Shepard for his part views it as nothing of the sort, but as a relatively recent historical development and as such – as yet – not shared by all children. To be sure, there are differences of emphasis involved between the two thinkers. For example, while Winnicott did focus on the way in which the teddy bear served to provide comfort in a situation where the child was experiencing separation anxiety when realizing his mother was not there for him, Winnicott's more profound interest was to explore the conditions for tolerating reality as independent such, including mother as an independent being, hence facilitating a breach with primary narcissism necessary for psychological maturity: the "good" object is that which survives the child's attempt at destroying it, wishing to do so because of the frustration to do with experiencing that the other is independent, hence not (not always, not exclusively) here-for-the-child. Winnicott's theoretical aim, then, is to demonstrate the connection between aggression and attempted destruction of the loved object so as to come to accept the reality principle (Freud) and the independence of others, that is to say, their being outside the control of the child and having needs, interests, and projects of their own. Learning how to use transitional objects is a part of this stage of development.

The interesting thing to note in the present context is that Winnicott has nothing to say about the role – be it actual and borne out, be it missing – of animals in the course of human development towards accepting reality and attaining psychological maturity. In fact, there is no mention of animals in Winnicott's most influential work, *Playing and Reality*. They do not figure as creatures encountered and explored, as fertile soil (sic) for the creativity and spontaneity, invention and imagination, of the child, at whatever age at that. Significant others only show up in the guise of fellow humans; the entities that may serve the function of transitional objects, crucial to reality-tolerance and to coping with separation anxiety, likewise are not animals but human-made toy pets (the exemplary teddy bear). And should an animal be part of this psychological universe, it would in all likelihood be a pet (a cat or a dog), not a wild animal.

I realize that pointing out the absence of animals in Winnicott's account of what he, to quote another of his book titles, called "the maturational processes and the facilitating environment", may not come across as a valid criticism. It will likely be regarded as external not internal to his project. Indeed, what did Winnicott miss out on in having nothing to say about human-animal relationships?

One reply would be: nothing of any importance. Why? Simply because the animals are largely gone, at least the wild ones; they are gone because the world in which they were richly present, abundant in types and numbers, is a world of the past. Today, children grow up in near-complete human-made and human-comprising environments, like it or not. So why attend to the role of animals if they are no longer present?

Shepard is only too well aware of this. For the first time in history, he observes, "animals are almost completely lacking in the child's experience" (1997: 143). It is only that, for him, this is the reason for writing about them, at this time. It is not a reason for passing them over.

The critical point I want to make vis-à-vis Winnicott is that, in portraying as universal a situation where animals have no important role to play in the psychological world (in a wide sense) of the child, Winnicott can be seen to exemplify what the "shifting baselines syndrome" I talked about above amounts to in modern developmental psychology: from a historical point of view, the lifeworld without animals that Winnicott's account takes for granted as the "normal" condition, is in fact a *degraded* one, a result of a long process of removing animals from the environment in which new generations of children grow up and take their bearings, and of turning those animals that are still left into increasingly domesticated ones treated as pets conducive to human needs, that is to say, as animals with little or nothing to show in terms of otherness in Shepard's sense.

To be fair, Winnicott wrote his major papers in the 1950s and 1960s and a lot has happened since then. Recent years have seen a return of animals in some parts of psychology, especially in clinical work and practice. Even though there is no mention of "animal-facilitated therapy" in Winnicott, it in fact began two centuries ago among the Quakers at York Retreat, an insane asylum in England, Shepard tells us. He also recalls Florence Nightingale's advocacy of pets for the chronically ill (1997: 147). To illuminate the present-day mindset, Shepard refers to Leo Bustad, a professor of veterinary medicine who thinks there should be wards in all hospitals where patients may keep pets, animal refugia from which the terminally ill may get animals on loan, professional referral systems, visitor animal friends in nursing homes, and so on. The list is seemingly endless when it comes to the usefulness of animals for purposes of therapy.

Unsurprisingly, Shepard's comments on this trend are scathing. To him, it has all the trademarks of utility: "the animal commodity dressed up as medical treatment", creating a situation where "in return for the work of a well-kept slave the animals in this bond get 'friendship', legal protection, food, and shelter" (1997: 149). The rhetoric of progress notwithstanding, "bonding" in the sense of an extended custody is not from primal human ecology; it does not help restore a kind of authentic human-animal relationship now about to go extinct. On the contrary, says Shepard, we "falsify our relationship to wild animals with our husbandmen's eye, social worker's agenda, veterinary tools, and breeder's agendas" (ibid.). There is nothing in the history of inter-species relationships to justify this new trend: "Animals were present at the center of human life for thousands of centuries before anyone thought of taking them captive, making them companions, forming the "friendship loops" of which animal-facilitation therapists and ethicists speak" (ibid.). The bottom line is that "wild animals are not our friends. They are uncompromisingly not us nor mindful of us, just as they differ among themselves" (1997: 151).

To corroborate his criticism of the current trend, Shepard seizes upon Mary Midgley's reference to "the 'human/animal bond', as if it were an ancient demonstration and right of amity" (1997: 150). To speak of a bond in this manner is deeply mistaken, Shepard maintains. Going back to our primal relationships with animals, we do find "a heritage of wisdom in the form of an ecological conscience, an attitude of humility toward the natural world, a sense of human limits and participation from a thousand centuries of the Pleistocene". However, this has nothing to do with "a display of entente cordiale nor an affectionate compact of familiarity or helping, centered on keeping, as if by a shepherd, nor a modern sentiment of gushing over kittens in a basket" (ibid.).

Is this a fair criticism of Midgley? I don't think so. There is no advocacy of human-animal bonding in *Animals and Why They Matter*, the book Shepard refers to. One would also look in vain for an argument to this effect in *Beast and Man*. That said, the former book does contain statements with which Shepard would no doubt disagree. Consider, by way of example, the following: "It is one of the special powers and graces of our species not to ignore others, but to draw in, domesticate and live with a great variety of other creatures. No other animal does so on anything like so large scale" (Midgley 1998: 111).

On Shepard's view, this "peculiar human talent" (Midgley) is a mixed blessing, to put it mildly. Humans' adeptness and ingenuity in "drawing in" and domesticating animals so as to live with them is a historical fact; and one can certainly be impressed by the ways in which humans have been able to find ever more uses for the animals that fall within the concentric circle. But Shepard's point is that this process comes with much too high a price – not only for the animals involved, but for the humans as well.

Along the spectrum from wild animals to pets, the result brought by ever-increasing domestication is that animals cease to be wild; they are turned into pets precisely insofar as, at the hands of humans, they have their wildness robbed. When the beast is tamed, it is a beast no longer. There is nothing "natural", in the sense of Aristotelian species-specific flourishing, to this transformation; pets, Shepard reminds us, are not part of human evolution or the biological context out of which our ecology comes. Rather, they are "civilized paraphernalia whose characteristic combination of accompaniment and accommodation is tangled in an ambiguous tyranny"; moreover, "the paradox of frenetic emotion and casual dismissal reveals our deep disappointment in the poet's ability to do something, to be something, that we cannot quite identify" (ibid.). When pets become surrogates for human companionships, or substitutes for the resolution of interpersonal social problems, so as to help their human owners cope with whatever hardships they experience, disappointment is bound to result, presumably to the frustration of both parties.

To be sure, the inclination to turn to animals as a way of coping with psychological problems such as anxiety, depression, or loneliness is not novel. What may be distinctly modern, though, is the perception of persons who turn away from human relationships and concentrate on their attachment to, say, specific animals, or to a particular place in nature, as exhibiting a behaviour of "compensation", that is to say, turning to the second best when the primary sort of relationship has broken down and as a result is given up on. The non-human objects are considered as sought out in desperation and for lack of the real thing, as a means of escape, ultimately a last resort of psychic survival, rather than as *bona fide* objects-cum-relationships that are appreciated as meaningful and valuable in their own right. Erich Fromm's statement in *The Sane Society* is a case in point, talking about the person who "tries to find security regressing to and identifying himself with nature, the world of plants and animals" (1955: 47; see Searles 1960: 196f.). As Freya Mathews observes, the sense of enchantment reported by persons who identify with non-human others as part of their overarching life experience and sense of identity and belongingness, refusing to put a premium on relationships with human others, is seen as "regressive" from the viewpoint of Western psychoanalysis. But, as Mathews comments, "to adopt this point of view is, of course, to beg the metaphysical question". Having grown up with lots of different animals around her, Mathews speaks about "the receptiveness that my animal familiars created in me, and it filled my whole being with a sense of being accompanied, of never being alone, a sense of background love, akin to the background radiation of which physicists speak. This is a love which has nothing to do with saving us from death and suffering, or with making us happy" (2016: 49f.).

I think that Shepard would concur with Mathews' point about (mainstream) psychoanalysis begging the metaphysical question as part of its failure to question the framework of human-centeredness. And although he accepts the criticism of overburdening animals-qua-pets with human expectations and needs they have no way of meeting, and so should be spared being dragged into in the first place, what he objects to is something more fundamental, namely the notion of "bonding". "Bonds", he argues, evoke images of babies and their mothers, of fraternal loyalty, whereas "interspecies bonds" implies something felt by both parties, albeit not in the form of strong reciprocity, where the same commitment is expected in return. Even so, Shepard maintains, a bond is not only a connection; there is a certain strength and depth to it, otherwise it doesn't qualify as a bond. Even more importantly, "bonds are legal instruments defining rights in terms of restraints on the freedom of the bonded to be themselves – that is, for animals, to be wild" (1997: 314). On this basis, Shepard observes that "it is difficult not to see in this reference to bonding by animal protectors a kind of barnyard hypocrisy. Am I to feel interspecies bonds to all other species? Am I bonded to bacteria as well as to horses?" (ibid.)

Notwithstanding the ad absurdum argument alluded to, rock bottom for Shepard is what he takes for a fact: namely that, "unfortunately for 'ethics', the natural community of life is not a bonding between species". Therefore, "to propose the tame animal and its owner as a model 'bond' for human/wild animal relationship is nonsense". Indeed, "it is as though all the virtuous and well-intentioned sense of obligation toward a few species of domesticated animals were seen as a guidepost for attitudes toward all animals everywhere" (ibid.). Shepard mentions the organization "Kids in Nature's Defense" (KIND), sponsored by the National Association for the Advancement of Human Education, whose pledge is "creatures great and small, we must defend them all". To him, this way of thinking shows that the notion of bonding has encouraged a dream where bonding has been taken into the wilderness "as though all life were kittens abandoned on the street. The clinical regimen for captives – therapy, care, and consoling – becomes the key concept. The world is to be a protected and prophylactic zoo, at once medical and moral, the object of charity and the anodyne for guilt" (1997: 315). Shepard looks upon this misplaced zeal as a symptom of the fear of death he considers to be deeply entrenched in modern Western culture. And as soon as the attitude to death is brought up in the context of humans' relationships with animals, that of the killing and eating of animals becomes a central, if not *the* central issue for discussion.

I shall postpone this issue, comprehensive and complex as it is, to later chapters.

As for the criticisms on Shepard's part that I have mentioned, I think that they succeed in seizing upon some of the crucial weaknesses in much of current thinking about "the ethics of animals". Shepard comes to the topic of animals and their place in Western culture with a depth of historical, ecological, and psychological knowledge that is rare, not only among his fellow philosophers. The weaknesses I think he is particularly good as exposing are not first and foremost of an ethical or normative kind. And to the extent that they are so, it follows from a failure to take the historical background as fully into account as Shepard does, allowing him a keen eye for the profound ways in which the human-animal relationship has changed, especially during the last century.

To conclude, Shepard's historically informed analysis helps us recognize how advocates of "interspecies bonding" represent yet another case of taking a degraded state of affairs as the normal one – taking for granted, that is, that animals are such creatures (or creatures in such a condition) as to allow "bonding" with humans to occur. That such a scenario, facilitating such a kind of relationship, should appear plausible, indeed as the most plausible one, to a present-day audience, is itself proof of how far down the historical trajectory from wild, over domesticated to tame animals we have come – though, again, with scant resources within the account given to identify and critically assess the changes involved. Again, losing the wild animal from view may certainly be held to be true to the empirical facts, the way things are these days – though on pain of showing up the transformation of animals – what they are, and how they appear to humans – that is the substantive core of the process.

2 Call and response, or eating and being eaten

I

In a central passage in her book *Interspecies Ethics*, the American philosopher Cynthia Willett gives the following quote from anthropologist Barbara Smuts' article "Encounters with Animals":

> Until recent times, all humans possessed profound familiarity with other creatures. Paleolithic hunters learned about the giant bear the same way the bear learned about them: through the intense concentration and fully aroused sense of a wild animal whose life hangs in the balance. Our ancestors' survival depended on an exquisite sensitivity to the subtle movements and nuanced communication of predators, prey, competitors, and all the animals whose keener sense of vision, smell, or hearing enhanced human apprehension of the world.
>
> (Smuts quoted in Willett 2014: 98)

Readers may be reminded of Shepard's historical account when Smuts highlights the importance of various animals' contribution to "enhance human apprehension of the world". In a manner typical of Willett's approach, she comments on Smuts' statement by raising the question of "how we modern humans might relearn this older knowledge" (ibid.), clearly holding this to be both a possibility and something desirable. "Animals, long ago vanquished from human society and consigned to the status of mute nature", writes Willett, "may once again participate as more than innocent objects of our pity or tropes of return to Edenic life. To varying degrees, they join in a multispecies social dynamic of call and response" (2014: 99).

We will see that call and response provide the model for Willett's interspecies ethics. Before discussing what this means, it is worthwhile dwelling on Smuts' work, acknowledged by Willett's as "a key source for rethinking traditional modes of ethical development" (2014: 207, fn. 1).

Smuts is doing research on primate social behaviour, concentrating on baboons living in Gombe National Park in Tanzania. Her reflection on the effect her experiences of the baboons came to have on her

DOI: 10.4324/9781003317692-3

is philosophically interesting, testifying to what is nothing short of a conversion:

> I felt like I was turning into a baboon. [...] I had gone from thinking about the world analytically to experiencing the world directly and intuitively. It was then that something long slumbering awoke inside me, a yearning to be in the world as my ancestors had done, as all creatures were designed to do by aeons of evolution. Learning to be more of an animal came easily as I let go of layers of thinking. This shift I experienced is well described by millennia of mystics but rarely acknowledged by scientists. Increasingly, my subjective consciousness seemed to merge with the group-mind of the baboons. [...] I had relinquished my separate self and slid into the ancient experience of belonging to a mobile community of fellow primates.
>
> (Smuts quoted in Willett 2014: 130, 131)

I cannot read this without thinking of Hans-Georg Gadamer's hermeneutics, with its warning against "going native": to abandon one's particular point of view in order to understand the epistemic object at hand (be it a text, a shaman or a baboon), and so to experience the *merging* of oneself with the object/the other, is not to attain the knowledge sought but to undermine it (Gadamer 1975). Only by holding on to one's point of departure as a ground for reaching out, for experiencing something as different, as representing a contrast and so facilitating friction, can the experience that is part and parcel of new insight be had. Not only does the way in which that object or other *differs* from me come into sharp relief that way; this holds for the discovery and appreciation of similarities as well. Representing different points of view, even embodying them, is not detrimental to understanding but the condition of its possibility. And it is only be exposing myself to what is not-me, to ways of understanding and of being in the world that are not mine and not those I have come to take for granted, as quasi self-evident and the only "natural" way, that I may realize that my (or my culture's) take on reality is just as selective, just as peculiar, as that of others. If the contrast that puts the other into sharp relief as not-me is lost, I lose not only the chance to learn something new about the other but the chance to learn something new about myself as well.

I shall have more to say about the importance of Gadamer's insight below. For her part, Willett is not troubled by the enthusiasm with which Smuts relates her experience of merging with the group-mind of the baboons. On the contrary, she concurs with Smuts in viewing it "as an example of the highest level of friendship attainable across species" (2014: 131), thereby challenging the proposition championed in "Hegel, Kant, Plato, and several thousand years of Neolithic metaphysics" that there is an abyss between human and nonhuman (ibid.). The legacy of this metaphysics, Willett contends, is that it diminishes what ethics could be. As against this, and at long last, historically speaking, "extraordinary discoveries from scientists are recovering a lost wisdom of interspecies friendships and communities" (ibid.). Smuts' research on baboons joins company with Jane Goodall's research on chimpanzees, Frans de Waal's on bonbos, and Marc Bekoff's on social carnivores to unfold what Willett describes as "a thickening plot of diverse ethical capacities across species and a renewed vision of the courtesy and limits required for common life" (2014: 132). Much in the spirit of Mary Midgley, Willett stresses the pioneering role of Darwin in this research, in particular his ground-breaking assumption that ethics originates in the affects, bonds, and meaningful existence of a social life. The exclusive valourization of linguistic and cognitive skills that has been the hallmark of the ethical theories in the Western canon mentioned above, forming the basis of their case for human exceptionalism, has come with the cost of losing sight of "complex biosocial attunements" of the sort that Darwin was able to identify where they were not supposed to be found – in non-human animals.

Key to the call and response model that Willett outlines is the role of "affect-laden, but not necessarily irrational or non-cognitive, attunement for the clusters or communal associations that compose a biosocial network" (2014: 12). Her claim is that "the unexpected agencies and communicative powers of creatures who may share overlapping habitats, histories, or communal structures with coevolving other species point beyond generosity or compassion to a more ordinary dynamic of 'call and response'" (2014: 13). In making her case, Willett relies on the research on affect attunement among human infants and adults carried out by the psychologist Daniel Stern in his pioneering study *The Interpersonal World of the Infant*. The call-and-response exchanges that Stern was able to observe in the interaction between a mother and her baby demonstrate that the mother can tune into and respond appropriately to the way the infant expresses himself or herself even though lacking adult cognition, verbal capacities, and – even – a preliminary sense of self, showing up the indispensable role of the mother in helping the baby develop a (proper, separate, distinct) self, facilitating its emergence by way of projecting its existence, to put it in a Hegelian way not alluded to by Willett. On Willett's view, Stern's research on humans can help us gain insight into the rhythmic, tonal, or gestural patterns of animal communication: "the [human] infant-adult exchanges point toward the possibility for communicative exchange across a significant range of nonhuman species" (2014: 14).

We should read this statement as programmatic. Willett's ambition is to use research on human infant-adult interaction as a basis for arguing the *extension* of the dynamics discovered to exchange across non-human species, and ultimately, to *exchange between humans and animals* such as reported by Smuts in her encounter with baboons. Hence, in a passage discussing Donna Haraway's work, Willett states that "the aim is to recognize animal agency and build intimate and political solidarities across species lines, suggesting an affirmative answer to Haraway's questions", these being: "Can animals play? Or work? And even, can I learn to play with this cat?" And further still: "What if work and play … open up the possibility of a mutual response?" (2014: 95; see Haraway 2008: 22).

I take it that the mutual response referred to is occurring between the cat and the human person who engages with it in play (work is a more tricky category, to be left aside for now). Let's grant that an animal such as a cat can play, both with other cats and with a human person; and that in both cases, mutual response is forthcoming: there is the call and response intrinsic to the activity of two participants engaged in playing with each other.

But not all animals are cats. Cats behave the way they do, be it towards other cats or towards human beings, following a long history of domestication; they are pets whose lives are inseparable from their coexistence with humans, not wild animals, animals encountered in the wild. Making distinctions such as this one between animals is important for a number of reasons: philosophical, ethical, and political. Applied to the aim that Willett identifies in Haraway's work, approvingly so: it is one thing to "recognize animal agency", understood as something traditionally denied them, not least by philosophers, with serious ethical consequences in terms of human-caused neglect, exploitation, and suffering. It is quite another – I shall argue – to proceed from that aim to the next one, that is, "to build intimate and political solidarities across species lines".

We need to be clear about what role the animals are thought to have in building the solidarities with other species – presumably including humans – spoken of. What capacities, following from the recognition of their agency that humans now are (or should be) prepared to grant them, would be required? What if it turns out that the capacities are of a sort found in some animals, but not in others? Would this be a problem insofar as *animals in general* seem to be the category that Willett has in mind?

Part of what I am getting at is that Willett wants animals to be active participants, not passive addressees, in the ethical and political arrangement she advocates as one of solidarity between different species. The scenario she considers as one of genuine progress would entail the empowerment of animals, allowing them to exhibit their full range of agency so as to make them genuine equals to their human counterparts within a "species blind" political forum, one where animals affected by some decision or policy are given a voice, one that will count as equally important to that of humans. Nothing less would be required of a species blind democracy that makes good on the ambition of cross-species solidarity.

But Willett does not endeavour to outline the sort of political arrangement that it would take to realize her ideas about "inter-species ethics" in a large-scale as opposed to small-scale setting. She sympathizes with Haraway's "feminist anarchic communitarianism" without offering her own political model. Haraway is credited with foregrounding the agency of animals: what they are capable of should be regarded as more important than what they cannot do, defined by way of comparing disfavourably with humans. So, even though Bentham's famous assertion that the question to ask about animals is not "Can they speak, or can they reason?", but instead "Can they suffer?", appealing to the virtue of pity in the face of animal suffering, represents an improvement over previous approaches, it nonetheless "misses the affective richness of a light-hearted encounter" (2014: 72). Much as animals' now-granted capacity for suffering warrants humans' pitying them, admitting animals (some) moral standing despite their being unable to speak and reason, a lot is still to be desired in terms of recognizing their full-blown powers of agency, that is to say, the qualities that permit them to come into their own as precisely agents (subjects), as distinct from objects; as creatures who do things, and do them their own way, rather than simply being at the receiving end of the actions of others. In keeping with this emphasis on agency, writes Willett, "Haraway's seductive attunement to the interspecies politics of camaraderie and play offers a delightful glimpse into an *autre*-ethics. Animals at play and anarchists at work (or vice versa) generate a vibe of synergy that can challenge social inequalities and hierarchies while serving to strengthen social bonds", pointing the way to "another ethics, one that enjoys glimpses of communion in play", described by Haraway as a "utopia of equality and spontaneous nature" (ibid.).

What this amounts to, inspired by Haraway's *When Species Meet*, is a "cosmopolitics of peace" where "polite encounters offer insight into trans-species politics". As Willett explains, "with or without what philosophers call logos, social carnivores such as wolves, humans, and other predators reveal themselves as capable of suspending aggressive sexual predatory drives to cultivate friendly relations with creatures who otherwise might have served as prey. The social dynamic of play provides an opening into what may well be the most egalitarian origin for ethics" (2014: 73). In the key function ascribed to it here, play needs to have its locus shifted from the interior psycho of the (individual, separate) subject to "a social arena that crosses species barriers" (ibid.). Where are such arenas to be found in our society? Willett answers by way of telling us that anecdotes of cross-species communion abound. Revealingly, her examples are taken from zoos, for instance one in Tokyo where "a rat snake named Aochan befriended a dwarf hamster named Gohan. The hamster was originally offered as a meal; Oachan refused to eat the animal and seemed to prefer sharing a cage with her; now Gohan even naps on Aochan's back" (2014: 74). Willett quotes Marc Bekoff, a biologist praised by Haraway for exemplifying "science undertaken as a mode of companionship" as opposed to the traditional impersonal spectator stance of detached observation: "Such pleasure-driven social capacities that 'our old Paleolithic brains' share with other species can reengage humans in the close inter-relationships with other beings that help us figure out who we are in the grand scheme of things". One can take this even further, Bekoff suggests, "to wonder if, when studying 'ecological interactions' – that is, encounters among different species of animals and interactions between animals and trees and plants – it makes more sense to concentrate on positive ecological interactions rather than on competition and predations" (ibid.).

I have cited Willett's example and Bekoff's statements extensively because I consider both to raise a number of important issues. To begin with, note that the cross-species exchange so central to Willett's project comes in quite different forms: there is the human-animal one exemplified in Smut's encounter with baboons; and there is the animal-animal one between a rat snake and a hamster. The latter exchange occurs within a zoo, and every step along the sequence between the two animals testifies to the unsaid in Willett's account: the presence of humans – be it in the form of offering the hamster to the snake as a meal, or in facilitating that the two animals share a cage with each other. As for Bekoff's comments, we recall that he practices science as "a mode of companionship", permitting him to ask big questions about how humans can reengage in the close relationships with other beings – crucially, animals – so as to figure out who we are in the grand scheme of things.

There are many questions worth asking here, for example: absent humans and the set-up they provide the two different animals in the Tokyo zoo, what would an encounter between them be like? In a sense, the question is moot: we have no way of telling, because no comparison is made between how their encounter plays out in the zoo, hence as facilitated, observed, and interpreted by humans and to humans, on the one hand, and their encounter in a human-free context, one the other. That said, most readers would probably hasten to add that, albeit factually lacking, the comparison does not really invite agnosticism: far from simply "not knowing", there is a strong assumption that, in the second scenario, the rat snake would have killed and eaten the hamster. Is it not this very assumption that explains the fascination expressed by Willett and Bekoff alike – namely that what we are able to observe between two different animals such as these is about "positive ecological interactions", rather than "competition and predation"? The fascination, that is to say, with what may come to pass across species once the individuals representing them get the chance to relate to each other in the form of play?

Let's take a closer look at Willett's analysis of play to appreciate what is at stake here. Like Berkoff, she considers that the enduring insight in Darwin's work is nothing to do with the "survival of the fittest" (in its vulgar reading) but with his truly revolutionary attentiveness to the emotional and social life of animals, making us realize that emotions evolved in animals no less than in humans for the purpose of furthering social bonds. In his important book *Descent of Man*, published in 1871, Darwin emphasized that, among social species, "the fittest" are not necessarily the strongest, nor the most aggressive, nor the cleverest, but the most sociable: those "whose temperament most inclines them to friendly cooperation", to quote Midgley's gloss on his position (2010: 49). Darwin's general conclusion is that "the difference in mind between man and the higher animals, great as it is, is certainly one of degree and not of kind" (Darwin quoted in Midgley 2010: 81). This conclusion, revolutionary at the time, followed compellingly from Darwin's own research, where he had seen that "the senses and intuitions, the various emotions and faculties, such as love, memory, attention, curiosity, imitation, reason etc. of which man boasts, may be found in an incipient, or even a well-developed condition, in the lower animals" (ibid.). In this late-Darwinian spirit, Berkoff considers that his studies of play behaviour in social carnivores show how it provides a training ground for learning the "Golden rule – do unto others as you would have them do unto you"; the animals he lists as cases in point are wolves, coyotes, red foxes, and domestic dogs.

What does Willett make of this? To her, play praxis – with its two key features, role reversal and self-handicapping (restraint) – contrasts sharply with "the abstract equality of atomized individuals before modern laws" in that "friendly play provides a training ground for a concretely situated reciprocity that binds selves-in-communities"; moreover, "codes of play protect the vulnerable by enforcing the requirement that the powerful set aside their power"; these, she says, are "the codes of anarchic communitarianism", that is, the kind of interspecies politics championed by Haraway (Willett 2014: 76).

This suffices to make the following point: in her use of the animal studies pioneered by Darwin and presently carried out by researchers like Bekoff, Willett passes from the animal behaviour being observed to the political arrangements of human societies, including institutions such as law. Analytically, two things are at work here. First, Willett clearly wants to invoke the animal studies findings for all they are worth as a weapon to critique powerful trends in modern society: atomism, dog-eat-dog competition, suppression of feelings understood as something inferior or irrational, egoism valourized as more rational and truer to our nature than altruism, etc. – these are so many prevalent notions about who we humans "are" and where we come from, in terms of our evolutionary trajectory, that ethological research helps reveal as so many prejudices, not only about us, but about our (allegedly profound) differences from animals. Secondly, the move that Willett relies on is from one object domain to another: from a context where animals and their behaviour are being studied, to a context where the agents whose behaviour matters are humans and humans only: namely contemporary society with its "abstract equality of atomized individuals", a society where bonds and solidarity – be they between humans, be they between humans and animals – are undermined by competitiveness and instrumentalization, making others in general either into enemies or mere means.

This is a commonplace critique of modern capitalist society that is not elucidated as such in Willett's book. I nevertheless take her to subscribe to it, if only implicitly, borne out in her critique of "liberal" ethical theories like Habermas' discourse ethics, and in her endorsement of the positions taken by theorists like Haraway and Judith Butler.

Here is where I see a main problem in Willett's work, to do with precisely the way she passes – glides – from one context to another: from observations concerning animals to a description of modern society, using the former for the purpose of critiquing the latter. I am not suggesting that "using animals to critique (human-centered) society" exhausts Willett's project, but it does figure as a prominent component of it.

To appreciate what I am getting at, consider this passage:

> In playful exchanges, wolves and other carnivores appear to break out of normal roles and social expectations, allowing creative and spontaneous responses to fellow creatures. The carnivore's rule – what humans conceptualize as the golden rule – is felt and formative, not formulaic. Self-handicapping fosters creative encounters where playmates differ significantly in capacities or desires and thus cannot be treated exactly like, or strictly analogous to how, one would want to be treated oneself. Determinate moral laws fails to account for this kind of flexibility because of their abstract formulation. [...] Social play exhibits an egalitarian element that goes beyond law's abstract equality. Participants in play treat each other as more than abstract equals. Through self-handicapping and role reversal, the dominant participant actually yields real and symbolic power to the playmate. This yielding of social and political capital grounds horizontal reciprocity and contrasts with the charitable giving of resources, which might display the character virtue of benevolence, but does not alter the social field.
>
> (2014: 76)

If this passage demonstrates how Willett moves from one context to another, it also attests to her tendency to make the move *in both directions*, as it were: not only, as we saw above, from animals to human society, but also from the latter to the former. There is no attempt to hide this on Willett's part: treating each other as "more than abstract equals", the "yielding of social and political capital", etc. – these are clearly behaviours that we commonly observe and comment upon in humans, taking it for granted that they apply there and will be meaningful not only to us as observers and commentators but to the agents whose behaviour we pronounce on as well (in the social sciences, this is the "double hermeneutics" of which Anthony Giddens (1984) speaks). Willett, however, departs from this employment of the terms; she uses them about animals, meaning them to be meaningful and valid as descriptions of their behaviour, exemplarily in play. The only animals mentioned in the passage cited are wolves, prompting the question whether the analysis is meant to apply to other sorts of animals as well, a question not raised by Willett.

What we realize, then, is that Willett is moving back and forth between animals and human society, allowing her to contrast animal behaviour with the norms, codes, and practices (such as morality and law) prevailing in modern society, and to employ concepts derived from the human-to-human-oriented practices to interpret the behaviour of animals, be it that they conform to the expectations that go with these concepts, be it that they deviate from them. And when they do appear to deviate, that deviation from "us humans" is largely painted as positive, lending the comparison the critical-subversive edge as against present-day political arrangements that I mentioned above. What is conveyed as positive in this regard is the way animals engaged in play are interpreted as expressing attitudes of friendship and egalitarianism, and a willingness to protect the vulnerable. Again, the reason why attitudes such as these are deemed positive is that they compare favourably with the "normal rules of hierarchical societies" (2014: 77); hence they are considered positive in a normative sense, showing up how animals can serve as moral models – a term never used by Willett but one I take to be in tune with her project.

I am not suggesting that what Willett is up to is a simple turning of the tables – replacing the conventional notion about human superiority over animals with its reversal. That said, there is no doubt that the assumption upon which Willett's interspecies ethics is premised is that we humans have something important to learn from animals: they teach us crucial lessons about life, understood as a profoundly inter- and cross-species phenomenon, not an intra-species one. Historically, however, the potential animals hold for humans in this respect has been largely suppressed, owing to the ideology-put-into-practice of human supremacy that has been hegemonic in the recent history of Western societies. Instead of being given a voice and the freedom to be what they truly are in their own right and of their own accord, animals have been put to use in so many ways by humans, and for humans, silencing their voices and turning their human exploiters deaf, dumb, and blind to their plight, and thereby to their true potential as companions. Thus animals have been denied the chance to be models of moral import – the very notion would be regarded as a *contradictio in adjecto*.

In my view, Willett puts animals to use no less than do the philosophers she wishes to reject on account of their anthropocentrism and hence – implicit or explicit – disregard for animals. The all-important difference would be that the way she uses animals is intended to grant them the full-blown recognition – moral and otherwise – hitherto denied them.

For all my sympathies with this project, I find that in Willett's case it is being carried out in a fashion that I find problematic for three reasons – methodological, epistemic, and moral – intimately related at that.

On pain of sounding as pedantic as I did when invoking Gadamer's hermeneutics to raise questions about Smuts' "going native" in her experience of baboons, I am critical of the explanatory role Willett ascribes to observations made of animals engaged in play, such as the wolves she mentions. Play is a particular type of interaction, raising questions about how representative the findings derived from studying it can be. The problem is that of wanting to portray as paradigmatic what comes to pass in a setting that is in fact highly selective. Midgley, always down to earth, makes a pertinent observation worth citing here: "Adults among other [than human] advanced species do indeed sometimes play together, and their doing so is rightly seen as a sign of intelligence. But they all do it far less than humans" (1998: 119). Secondly, to the extent that Willett's empirical material is drawn from what she presents as "playful exchanges of wolves and other carnivores", the behaviour observed is inter-species, but not cross-species. To be sure, we do learn to appreciate the kinds of social bonds, based on and sustaining social-cum-moral emotions such as concern and sympathy, that such animals as wolves exhibit in the way they engage in "call and response" with each other – meaning with their own kin – be it playfully at that. But a more important feature of the behaviour of wolves is to do with their being – precisely – carnivores: they are predators that depend on prey, that is to say, on seeking out other animals for the sake of killing them. Doing so requires a quite different set of behaviours than that involved in play, raising the question of what part of the repertoire to regard as at the core of being a wolf. It is telling that in the course of Willett's discussion, there is no mention of the role that "eating and being eaten" (Shepard), killing in order to stay alive, plays in the lives of animals like wolves. Indeed, death is conspicuous by its absence in Willett's approach, save for in connection with a discussion of how the main character in John Coetzee's novel *Disgrace* experiences a conversion in his attitude to the (in)human practice of killing animals in the industrialized fashion currently dominant, this being a very different context for discussing the reality of death than the one I have in mind, i.e., what animals do to each other as predator and prey.

One thing that Willett does dwell on, however, is the well-known claim made by Heidegger to the effect that whereas animals simply live, human persons are unique in their awareness of death; indeed, one person's "being-toward-death" (Heidegger) is an individualizing feature, related to as a generic one only on pain of inauthenticity. To challenge this notion of human uniqueness with regard to death, Willett asks us to consider Jane Goodall's account of "the chimpanzee mother who cannot abandon her dead infant and who carries the body for days", prompting her to ask: "Could chimpanzees and whales be people too?" (2014: 119).

Why does Willett raise the question like this? Why is it important that we come to see animals like chimpanzees as people too, or at least as more similar to humans than we have used to do? Presumably, doing so would be crucial in admitting such animals a standing more equal to that accorded to humans. That is why extending, say, the capacity to be aware of one's own death, or the capacity to mourn dead loved ones (like the chimpanzee mother did) to animals, and so depriving these capacities of their long-alleged human uniqueness, would be welcome from the point of view of doing justice to the animals as our moral equals. So the formula would be: the more animals come to be seen as sharing (some of) the capacities we humans have taken to be unique to us, the more they should be seen as qualified for inclusion in the moral universe. In other words: Humans' discrimination of animals has been based on assumptions about their differences from us; when these differences turn out to be non-existent – one after the other, tied to capacities – the justification for treating them as inferior would fall away.

Before I consider the implications of this reasoning, a closer look at Willett's reference to Goodall is in order.

As we saw, Midgley is just as interested in Goodall's work as Willett. At one point in *Animals and Why They Matter*, Midgley mentions Goodall's account of the effect of a polio epidemic on the chimpanzees she was studying. She explains:

> One old animal, his legs wholly paralyzed by the disease, was dragging himself around with his arms. He was suffering from loneliness, since he, like the other crippled individuals, was shunned and sometimes attacked by those who were still healthy. In the hope of inducing two companions who were grooming each other to groom him as well, he dragged himself up into a three. [Goodall writes:] "With a loud grunt of pleasure he reached a hand towards them in greeting – but even before he made contact they both swung quickly away and, without a backward glance, started grooming on the far side of the tree. For a full two minutes, old Gregor sat motionless, staring after them. And then he laboriously lowered himself to the ground. As I watched him sitting there alone, my vision blurred, and when I looked up at the groomers in the tree I came nearer to hating a chimpanzee than I have ever done before or since".
>
> (Midgley 1998: 120f.; see Goodall 1971: 202)

We should note here, as does Midgley, that Goodall knows perfectly well that it would be "unfair" to hate the chimpanzees on the basis of their behaviour towards Gregor, "since they are incapable of reacting otherwise" (Midgley 1998: 121). In fact, Midgley's interest in the episode is more to do with what it says about humans than what it says about the animals: what is demonstrated in Goodall's reaction ("I came nearer to hating a chimpanzee") is "the wider power of human sympathy" as compared with that found among animals like chimpanzees. Only occasionally will chimpanzees help a sick comrade instead of being repelled and retreating, whereas wolves, elephants, dolphins, and whales are known to do it rather more often. The point Midgley uses Goodall's account to make, then, is that human beings have "a much wider capacity still" for extending sympathy to others not limited to one's family group, a capacity which is perfectly able to cross the species barrier. In other words, the emotional preference for one's own species over others seems to be prevalent among animals including higher mammals, who – for all that, as we see in Goodall's example – may abandon rather than care for what they perceive as a sick or dying member of their own species. By contrast, finds Midgley, humans are quite well equipped to notice and respond to the moods and needs of other species – which is not to contend that human persons will choose to act on this capacity for sympathy with, say, a sick animal whenever such a situation arises. Rather, the point is that were the person *not* to act like that, his or her not doing so would be met with criticism, since regarded as a sort of moral failure – precisely the reaction that we saw is *not* appropriate if directed at animals like chimpanzees.

I realize that this way of looking at it may come across as perfectly conventional, perhaps disappointingly so. To be sure, the moral dimension hinted at in speaking of a given response as *criticizable*, and understood as a prerogative of humans, is a complex issue in its own right, one I shall come back to in the next chapter.

II

The episode with chimpanzees Midgley cites from Goodall's work differs markedly from the episodes that Willett focuses on. Chimpanzees, evidently, are capable of quite different sorts of behaviour: in one case, there is the mother who cannot abandon her dead infant and who carries the body for days, inspiring Willett's question: "Could chimpanzees and whales be people too?", a question that I take Willett to indicate to her readers should be answered in the affirmative – or, to put it more cautiously, a positive answer is something we should seriously consider rather than dismiss out of hand. However, with regard to the episode discussed by Midgley, the question seems to invite the opposite answer: that in situations where a response like sympathy becomes an issue, chimpanzees will act so differently from humans (or from what humans expect of themselves) that the conventional view about how they differ from us humans is confirmed.

I find Willett's presentation of the animals studies she draws upon tendentious in concentrating too one-sidedly on material where what readers are struck by is the pro-social, even pro-moral, features in animals' behaviour, especially as brought out in play. I am not saying that these features do not exist in the animals in question. What I am suggesting is that they only help us see half of the picture, ignoring the contexts and situations where very different behaviours are on display. From a methodological point of view, the selectivity that we encounter in Willett's discussion is to a large part due to the selectivity of play, this being her preferred case, and hardly coincidentally so in that it does seem to offer empirical evidence of a sort that corroborates her claims about pro-social behaviour among animals, even towards members of other (animal) species.

However, the methodological problem runs deeper: it is not limited to the systematic preference of activities like play over activities like hunting; a selectivity that must be considered especially problematic when the animals in question are carnivores like wolves – can their distinct way of being-an-animal really be described when the features they require to kill in order to sustain themselves qua wolves is not accounted for? That said, the deeper problem I see is brought out in examples like the one we looked at above where "a rat snake named Aochan befriended a dwarf hamster named Gohan". I already mentioned that it must be considered significant that the interaction that occurs here does so between two animals in a zoo.

This prompts the following question: What would the interaction between these animals be like if it took place not in a zoo, but in their natural habitat? As I indicated earlier, there is a sense in which Willett's presentation trades on this contrast: she implicitly acknowledges that we would expect the rat snake to attack the hamster, only it turns out in the example given from the zoo that it does nothing of the sort – instead, the rat snake befriends the hamster. The surprise Willett takes this to amount to derives from the presumed expectation that the interaction would be about attack rather than making friends.

If we grant, for the sake of argument, that an encounter between the two animals in their natural habitat would exhibit a very different sort of behaviour than that reported from their encounter in a zoo, which of the two instances should be regarded as the one most true to the facts? Should we consider that the one context is "natural" and the other "artificial", implying that the observations accessed in the former case are more reliable, scientifically speaking, than the latter? Or should we rather say that the natural/artificial distinction is untenable, that it no longer holds? Why would that be? Well, because those days are gone when animals who used to live off limits from contact with humans do so; their environment is becoming impacted by the activities and presence of humans to such an extent that the notion of a "natural" habitat as a kind of "separate" domain, one where the "true" nature of the animals in question is on display, is becoming obsolete; so that for lack of being intact in its original sense, the very notion should be discarded. Indeed, are we not already, as of today, in a situation where animals live their whole lives in an arrangement with humans, having their environment shaped by humans, be it intentionally and by design (i.e., in a zoo) or not, so that the behaviour they now exhibit in that normal setting is to be regarded as their "natural" behaviour?

True, these are big questions, pointing beyond Willett's book and to be more fully addressed in Chapter 5. One thing to note here, though, with the example just discussed in mind, is that the way I just used the natural/artificial distinction is different from being concerned with the ways in which, say, zoo personnel – or for that matter, Smuts or Goodall – can be said to impact, by way of their sheer presence, the animal behaviour being observed. Trying to measure or assess the significance of "human impact" on what comes to pass is notoriously difficult; nowadays many scholars would hold that the very exercise of "isolating" or "detracting" the impact of humans is based on an untenable assumption about a clear-cut division (demarcation) between the (human) epistemic subject and the object. Otherwise put, what is at stake is not the impact of humans on the microlevel, in one-to-one encounters (episodes) with animals, be it in a zoo in London or in big national parks such as Serengeti; rather, what is at stake is the historically unprecedented situation created by (in) the Anthropocene, where everything that comes to pass does so in a way owing to, shaped by, the impact of humans, be it directly on the scene or be it in more subtle and less visual-physical ways difficult to outline and identify, in time and in space, but decisive nonetheless. It marks a limit to Willett's discussion that she does not engage with these larger questions.

To conclude, consider this passage where Willett summarizes her findings:

> Extraordinary discoveries from scientists are recovering a lost wisdom of interspecies friendships and communities. Smuts' research on baboons joins with Jane Goodall's research on chimpanzees, Frans de Waals' on bonobos, and Marc Bekoff's on social carnivores to unfold a thickening plot of diverse ethical capacities across species and a renewed vision of the courtesy and limits required for common life.
> (2014: 132)

Let me put it like this: to support her case for an *inter*-species ethics, the research Willett draws upon would need to illuminate how animals behave when interacting with animals from other species than their own. The example with the rat snake and the hamster fits this bill, though as we have seen it raises a number of critical questions in its own right that Willett fails to address. As for the research listed by Willett in the passage cited, its shared main focus is on how members of one species (say, chimpanzees) behave towards each other. But studies of *intra*-species interactions do not *per se* qualify to offer us insight into what goes on in *inter*-species interactions.

Above I mentioned the problems arising when passing from one context to another – e.g., from natural habitat to human-made environment, or from what conservation biologists term *in situ* and *ex situ*, respectively. The problem I call attention to now is a different yet related one, more immanent to Willett's project: namely, passing from *one* species and the behaviours studied there to how interactions *between* different species of animals play out. In this regard, Willett proposes a model for an interspecies ethics that aims to "tease out layers of ethical engagement within and across animal societies". She goes on to speak about "capacities for cross-modal communication across differences of species and rank"; drawing upon these capacities, "biosocial norms and expectations for reciprocity and distributed responsibility bind together members of more or less cooperative societies", allowing us to recognize how "social rituals such as reconciliation and forgiveness may also on occasion cross a species barrier, altering the social climate and revitalizing life" (2014: 132).

What needs scrutiny is the way in which Willett takes the research she invokes to corroborate her claims about capacities conducive to the ethical engagement she postulates "across animal societies". To be clear:

there may be such capacities; if they exist as Willett suggests, they are vitally important. The problem is that we cannot really know, based on the research she draws upon and the way she presents its findings. In the statements I just quoted there is a thoroughgoing ambiguity as to who Willett has in mind when talking about "cross-modal communication", going on to mention phenomena of a social and moral quality that seem to be successively more demanding in terms of the capacities required: reciprocity, distributed responsibility, and social rituals such as reconciliation and forgiveness.

Who is engaged in reciprocity with *whom*, who assumes responsibility for whom, who takes part in the rituals of reconciliation and forgiveness, with whom as their counterparts? Do all sorts of animals take part in these practices, or only some in some of them? We are not told, not in any of the instances referred to. The description offered is so general in its aim to say something that is meant to hold for all animal species, or for a great many of them, that the result is a complete lack of specification in terms of the species involved. It is as if they can all be described in like manner, showing up the same (elementary) capacities for affects and bonds, helping bring about the more demanding, complex, and sophisticated social and moral phenomena just reviewed.

And yet, in some instances of inter- and cross-species interactions Willett does tell us who the participants are. Apart from the case of the rat snake and hamster and a few others like it, we hear about how one party in particular experiences what comes to pass: the human one. There is Goodall's fascination with the chimpanzee mother who carries her dead infant for days, and Smuts feeling that she was "turning into a baboon", and the like. Unfortunately, these experiences on the part of the scholars seem to be beside the point Willett's sets out to make. Why? Because we are left in the dark about how the situation being portrayed from the scholar's point of view is experienced by the animals – and it is the animals' way of life and of experiencing things – including various others – that we are concerned with, is it not, rather than how humans look at animals? There is no reason to question the authenticity of the psychological experience that grips Smuts as she warms to the baboons, made possible – perhaps – by their warming to her, getting used to her presence. But is the authenticity of the human person reporting about his or her reactions the appropriate validity claim here? In that case, my earlier Gadamer-inspired critique of Smuts will come across as old-fashioned, even positivistic, as wedded to a model of knowledge-seeking at a distance from its object that would be mistaken in the study of animals; a model, that is to say, where "scientists working with animals make considerable effort not to be seen by them", such that "good scientists", in Haraway's formulation – and she means to critique this model, not endorse it – "are those who, learning to be in visible themselves, could see the scene of nature close up, as if through a peep-hole" (2008: 24; see Despret 2022: 552).

Regrettably there is no discussion in Willett to inform us about her views on this matter. That being so, and setting aside for now the irremovable fact that what can be said about animals will be so from a human point of view (one way or the other), *the animals*' side to the encounter oddly fades from view. What one is left with is the kind of responses – predominantly emotional in Willett's material – that animals' behaviour, be it directed at other animals or at the humans present on the scene, or observing it from the outside – elicit on the part of the humans. Again, what does this tell us about the animals?

Above I spoke about what I called the formula according to which humans' discrimination of animals has been based on assumptions about their difference from us, the implication being that the more evidence we come to accept to the effect that these assumptions are wrong, the better. If animals should turn out to be more similar to us in terms of capacities and qualities we deem relevant – whereby those we regard as relevant to assessing humans must be admitted as being so in assessing animals as well – then this dismissal of one "difference" after the other must entail that discrimination loses its justification: the facts simply don't vindicate it, but instead cast it as unwarranted, arbitrary and downright immoral.

To be sure, Willett is correct in holding that Darwin paved the way for the then-revolutionary insight that ethics originates in affects and bonds that are not unique to humans but found in animals as well, to put the point simply and in a fashion alluding to the "moral sentiments" theories dating back to Hume, Smith, and Hutchinson. A better scientific grasp of various capacities, be they emotional or cognitive, has forced an entire series of long-held assumptions about human-animal differences to be rejected as simply not true to the facts as we now know them.

However, as I indicated, there is a catch to this reasoning if what it amounts to is the following formula: *the more the animals appear to be like us, the better* – better taken in a normative sense to do with implications for moral status. What is the problem?

The problem is that humans serve as the model, approximation to which on the part of (some) animals will be rewarded by their enhanced status, rightly so scientifically speaking and deservedly so morally speaking. We see that there is not one anthropocentrism, but two: one is the position that humans are unique and that everything that goes to demonstrate how animals differ from us testifies to (proves) their inferiority. This version of anthropocentrism was always two-in-one, normative as well as descriptive. The other version of anthropocentrism is the one I am getting at with my present argument: the claim about human uniqueness is given up, but it is so in such a manner that specific animals' now-acknowledged degree of *similarity with humans* serves as the parameter for determining their status and how they are entitled to be treated: the more intelligent (like we humans are) and the more capable of sentience (ditto) they appear, the "higher" their score; they are allowed to ascend the ladder with us humans at placed at the top, but the thing is: they are getting closer to us than ever before. Note that in this perspective there is no standard involved that derives from what is *peculiar about animals* (take birds' capacity for flying, something humans cannot), peculiar in a sense that would go to show that they (say, birds) are superior to us (they can fly). Such a shift of ground for making the comparison is precisely what is not undertaken, not entertained: we persist as the point of reference and comparison from which we never let animals escape. *Homo mensura* remains unchallenged and hence intact in this foundational sense.

This is an important point in its own right and one that I will develop further below. As for Willett, I suppose that if the formula I stated above were put to her to comment on, she would say she disagrees with it – at least in its most blunt form: the more the animals appear to be like us, the better as far as their morally relevant standing is concerned. Willett would not sign up to such a position, owing to its undeniable anthropocentrism. But to want to distance oneself from the *homo mensura* framework, is one thing; to actually show how her proposed model for interspecies ethics breaks with it and offers a convincing alternative to it, is another – and I don't see Willett doing that.

III

The Australia-based anthropologist and philosopher Deborah Bird Rose begins her book *Wild Dog Dreaming: Love and Extinction* by relating how one day she was told by a friend that "something awful" had happened: not far from Canberra she had seen a tree that was strung up with dead dingoes. When Rose went there to see for herself, the horror reported by her friend was confirmed: "the dingoes were suspended by their hind legs, heads down, bodies extended, another 'strange fruit' in the annals of cruelty" (Rose 2011: 1).

Rose goes on to say that many of her Aboriginal teachers, especially Old Tim Yilngayarri, had told her wonderful stories about dingoes, typically ending on the note "you've gotta leave him; no more killing". The context for speaking like this is a situation where dingoes are regularly poisoned and shot; indeed, "the shadow of death falls heavily upon their future". To be sure, "they are not the first animal to be facing extinction, and they will not be the last. But they are one of the few whose extinction is actively being sought by some segments of human society" (ibid.), that is to say, present-day Australia.

Although Rose covers much of the same territory in her book as Willett in hers, this opening passage signals that Rose's tenor is very different from Willett's: whereas there is no emphasis on death in Willett's account, Rose cannot talk about animals without addressing their extinction, understood not as a philosophical issue or a possibility in the future, but as the truth about their present plight. The extinction Rose has in mind, we should note from the outset, is one that is not "natural" but is instead caused by humans, sometimes to the point of being actively pursued.

To address extinction is of course to be concerned with death, though – yet again – with a death that is not the other side of life, and that has always and will always be so, but instead with death as brought about by killing performed by humans, as opposed to hunting performed by them as well as the death caused by other animals in their natural habitat. There are many ways to die, also for animals, and the differences between them are vastly important, especially in terms of what they reveal about the relationship between killers and killed. In speaking about killing in what follows, I mean acts done by humans, intentionally at that.

Why would (some) humans want to kill dingoes? Is something amiss when they do so? How does it reflect on the human-animal relationship, considering that such killings have become more frequent than in the past?

To get to grips with these questions, we need to know more about the dingoes. As we saw, Old Tim is Rose's teacher and guide, born in 1905 and a resident of Victoria River, one of the great monsoonal rivers of North Australia. Dingoes are the wild dogs of Australia, and Old Tim had a special relationship with them throughout his life. The dingoes figure prominently in the mythology of the Aboriginal people in the region; but unlike many Indigenous, which tell of specific places and specific people, the dingo stories told by Old Tim are meant to concern all human beings. The premise is that animals and humans have a common origin and destiny. In Tim's stories, the dingoes are cast as the ancestors of *all* human beings; "they give us our faces, our stance, our death, and the return that cycles us through the bodies of other living creatures. At first there was only one creature – a dog-human person – and this creature differentiated himself/herself, inaugurating both dog-persons and human-persons. Dogs/dingoes are still close kin because of their shared origins" (Rose 2011: 7). Such a differentiation is a commonplace in many indigenous cultures and will be familiar to readers of ethnographic studies from regions in the Americas as well as in Asia and Oceania (see Descola 2013; Kohn 2013; Viveiros de Castro 2015). Rose goes on to point at that Old Tim universalized this story about a common origin of animals and humans: "Dreamers [creators] worked that way for everyone, White Lady, Aboriginal men, all the same. They walk, they stand up, they're finished being dogs now, they're proper humans, women and men. Mother and Father Dingo made Aboriginal people. White children come out of a white dog" (ibid.). Rose urges us to think about it: "to look at the face of a dog is to see your own ancestor and your contemporary kin. It is to see mothers and fathers, sisters and brothers" (ibid.).

Evidently, and horrifyingly, this sense of common origin and of being kin is no longer experienced by the humans who nowadays kill dingoes and hang their bodies up in trees, as Rose saw for herself. What happened? Has the experience of kinship, of being brother and sisters and so having the same parents, simply been lost along the way, as time has moved on and there have been major changes in the relationships between animals and humans? Have the old stories of old people like Tim lost their relevance to younger generations? Has there been a shift in values, and hence in attitudes, towards animals like the dingoes in recent decades? What does a person who kills a dingo the way Rose describes it think about his (yes, we are talking about men, not women) doing so? What sort of a statement is such killing? And how do other people react?

Reading Rose's book, I had difficulties getting to the bottom of these killings. We are told about some of the factors that appear relevant to the phenomenon, but much remains unsaid. Perhaps it's too painful, so much so for Rose, that hints must suffice, disappointing the desire for a fuller explanation.

Even so, death has a strong presence in Rose's perspective, as a philosophical issue, an ecological one and an ethical one. The dingoes play a crucial role in enabling Rose to render concrete and particular what would otherwise be a general discussion about death and extinction, common to most philosophers' treatment of both. That *life depends on death* is a fact, however uncomfortable. Facing it as such has never been easy, not for any people or culture that we know of, and yet it seems that contemporary society is increasingly uncomfortable with death as the flip-side of life, as that without which life is not to be had. Historically, the main trend has been all about the removal of death from sight, obeying the "out of sight, out of mind" formula; this holds not only for humans but for animals as well – whereby, as I shall discuss in greater detail below, crucial differences may be observed between different animals: we handle our deceased cat or dog over to the veterinarian, or sometimes choose to bury our dead pets in the garden, but we don't want to see, hear or know – exactly – about how the pigs and chicken that we put on our dinner plates have come to die so as to end up there – have died by being killed by us humans, at that, in industrialized factory fashion in our days, at that. Again, we are reminded that some deaths are the result of killing, raising questions that we find uncomfortable.

Where should we locate the dingoes in this picture? They do not seem to fit either of the categories just deployed; they are not pets and part of the family at home, nor are they stored away for industrial processing (including highly mechanized killing) in factories whose inside most of us have never seen and never will. But they are not living in the wild like wolves or grizzly bears either. "Wild dogs" in Rose's characterization, they are used to the party of humans but don't depend on living with them. Old Tim, we recall, spent his whole life sustaining a relationship with some of them, allowing for a sort of bonding with them true to the spirit of the myth of common origin. Presumably the individuals who these days either shoot the dingoes as depicted above, or who take part in a systematic – and declaredly legal – campaign to have them killed by poisoning them, are not recruited among those who have built one-to-one relationships with them like Tim did.

If the stories told by Tim and highlighting the myth of a common animal-human origin now cease to appear relevant, this is a sign that the cosmology lending them meaning and plausibility is in the process of losing its grip on people's minds. The cosmology of which Tim is a late embodiment, explains Rose, is a form of *animism*, committed to the acted-upon view that "the world is full of persons, only some of whom are human, and that life is always lived in relationships with others" (2011: 18). In other words, the world as experienced and negotiated in various practices such as hunting is a world consisting of subjects not objects: humans always and everywhere encounter a someone, a *who*, within a rich and heterogeneous multitude of such, and not a something, a *what*; a Thou in some guise or another, not an It (to use Martin Buber's distinction). Within this context of what anthropologist Eduardo Kohn (2013) calls an "ecology of selves", of so many types of selves sharing the same place, the same territory, each one of them is only able to thrive and to sustain itself as long as all its others are allowed to play their roles in the totality everyone relies upon. As we saw, this is not all about love and care and building bonds, at least not in the modern sense of an affirmation of life's positivity to the exclusion of life's negativity, meaning death. "In a world of hunting and gathering", states Rose, "death and continuity are core aspects of the integrity of life, and are always unavoidably present in people's lives and minds" (2011: 18). Here, "an ethical response to the call of others does not hinge on killing or not killing. It hinges on taking responsibility for one's actions. Responsibilities are complexly situated in time and place; most of all they are up-close; face-to-face in both life and death" (ibid.).

Call and response, then, are very much at the centre of Rose's perspective, as they are in Willett's. But whereas Willett one-sidedly situates the phenomenon in play and casts it in a joyful and harmonious light foregrounding the "comradeship" between those involved, Rose is keenly attentive to the presence – as possibility or actuality – of death in the exchanges that occur, be they between animals or between humans and animals, exemplarily in hunting, a practice not discussed in Willett.

To recognize the centrality of death is to show how it is a topic at the heart both of a cosmology like animism and of ecology, understood as a modern science. Referencing the influential book *What is Life?* by Lynn Margulis and Dorion Sagan, Rose reminds us that life on Earth is only to be had as a shifting balance that is far from equilibrium and thus precarious. Humans emerged in dynamic relationships with animals and plants; "with them we share our dependence on water and air; and we share basic energy and basic substance", so that for us humans, understanding how we fit into the community of life and death that we share with so many different others is not an optional extra: as Rose's colleague Thom van Dooren puts it, we are "interwoven into a system in which we live and die *with* others, live and die *for* others" (2011: 20).

To sustain connections across species, then, means to be involved in sustaining them across death – again, life can only persist that way, with death as its *conditio sine qua non*. This is the theoretical and general way of putting it, in keeping with what ecology as a scientific discipline teaches. Rose's project is more concrete: it is thanks to encounters with persons like Old Tim and Aboriginal people that she has been able to bring the death narrative that is key to their (animist) cosmology into ecological domains, as exemplified by the current plight of the dingoes. In this context, "country" is the core concept; it means not only a spatial unit, but also a narrative about how the Australian continent is crisscrossed with the tracks of the creator beings, called "Dreamings" in Aboriginal English. Dreamings would perform rituals, they would leave parts or essences of themselves; they would "look back in sorrow, and then continue traveling, changing languages, changing songs, changing identity" (2011: 17f.). The latter is particularly important, pointing to the phenomenon of metamorphosis that is so crucial in many indigenous cosmologies around the world, that is, how animals may change into humans, humans into animals, back and forth (see Viveiros de Castro 2015: 220ff.).

Rose's presentation is worth citing in extenso:

> From the perspective of country, a death narrative binds the living and the dead into an ecological community (not just a historical community), working with multispecies crossovers as well as generational crossovers. When death is embedded within a system of multispecies kinship, animals that die are members of families; they have human kin as well as kin of their own species. In these Aboriginal families there are many deaths, not only because this is how life always ends, but because these people are hunters. Their own lives depend on the food that is the bodies of other animals, and every hunted animal is someone's family. Death is extremely intimate, and all deaths matter.
>
> (2011: 21)

Given this account, we are in a position to realize what is amiss in the current situation where dingoes are killed in the ways described above. We realize that this is just as much about *how* they are killed as about the fact *that* they are killed. They are not killed as part of the practice of hunting, understood as based on the necessity of humans' eating animals in order to survive, forcing the humans involved to recognize their utter dependence on the animals, a recognition that entails attitudes of awe and respect and feelings of gratitude and humility on the part of the humans, meaning both the individuals who participate directly, their families, and the whole community. Rituals see to it that the hunters take care to never take out more animals than dictated by the need for sustenance and survival; to be dependent on the animals' "giving themselves up", as a gift, means to regard the animals as the superior party, in that *humans need them more than they need humans* (see Berkes 2008: 103). Hence excessive or careless hunting is condemned by the community as a violation of the bond between humans and wild animals, putting into jeopardy the willingness of the animals to sacrifice themselves in the future, thus backfiring on the humans relying on hunting in that area.

There is a vast ethnographic literature on hunting that I shall not go into here, nor does Rose in her book. The points she makes at this stage are philosophical, highlighting the sharp contrast with our modern Western tradition that emerges here, a tradition that has been intent on finding ways to turn our eyes away from the deaths of animals, robbing their death of the intimacy and meaningfulness granted them in indigenous cosmologies, reminding humans of their connectedness with and dependence on the animals. In particular, Rose (using Val Plumwood's term) mentions the "hyperseparation" maintained by dualism á la Descartes, positing a categorical and absolute divide between humans understood as creatures with minds, intelligence, volition, and feelings, on the one hand, and animals (in general, taken as one homogeneous category) understood as "mere" nature (Descartes: "automata"), on the other, with the entire series of binary oppositions that correspond with it: humans as active, animals as passive; humans as subjects, animals as objects, etc. And although much of twentieth-century philosophy has been dedicated to critiquing the Cartesian legacy in a number of areas, as we saw earlier the sharp divide between humans and animals with regard to death is reiterated by prominent thinkers like Heidegger: only humans truly exist and possess the capacity to relate to their own situation, to their past, present, and future, hence only human death is an event, carries significance, and is of ethical import. By contrast, when animals die it is a case of "mere death" and as such a "fact of nature" understood as a realm devoid of meaning and value, agency and subjectivity, intelligence and volition. Human exceptionalism á la Heidegger, and for that matter Sartre, kills two birds with one stone, as it were, asserting the superiority of humans by way of stressing the inferiority of animals, rendering them passive objects that humans can kill with impunity. In this way, the death of animals is removed from the ecological as well as symbolic setting of multispecies and generational crossovers that indigenous cosmologies perceive them – animals, their life and their death – as inseparable from.

A death that is not indifferent "because" it is that of an animal and not of a human, is an event that as such is not only, or even primarily, negative. On the contrary, the death of an animal has its own peculiar beauty; that it does so means that it is something that calls for the appropriate attitude, that of wonder and respect, as were it something holy, although it – death – is precisely *diesseits*, not *jenseits*, a part of this world, not a beyond.

In this perspective, Rose does not shy away from describing her own experience with the death of animals – be it at her own hands – in outright poetic terms, something that would amount to a category mistake in the Western philosopher giants just mentioned. This is her depiction:

> The beauty of death lies in its mystery. To see the light of life leave the eyes of a dying creature is to see briefly into a region that is unknown and unknowable. [...] I learned to see the light disappear from the eyes of animals when I went hunting. Often an animal would look at us in the moment before it was shot. We saw it die; sometimes we had to finish it off with our bare hands. And then there was the distribution. When you cut into an animal, the thing that is so surprising is the reminder of how warm mammals are on the inside. You have your hands inside the animal, and you know without any doubt that the way this animal feels to your hands is exactly how you would feel if someone were doing this to you – the same heat, the same textures, the fresh smell, the red blood. That intimacy of interchangeable interiority forms a special kind of empathy based on the tactile knowledge of our mammalian kinship and our shared condition as creatures born to die. This dead animal could be me, and I myself will one day be a dead animal. [...] within these tactile webs of immediacy, killing is part of life because death is part of life. To be alive is to know that one's life is dependent on the deaths of others.
>
> (2011: 26)

We could have found this depiction in Shepard, but we do not find it in Willett. My impression is that the large majority of Western scholars in the fields of environmental philosophy and environmental humanities, as well as animal studies, oppose hunting, condemning it as unethical and as a sort of practice that belongs to the past and that they themselves would never participate in. Within this group, many will be vegetarians or vegans, based on a moral conviction about the inviolability of animals, seeing it as part and parcel of their moral status. In this way, being against the sort of killing of animals that takes place in hunting is to follow through on a principled rejection of human exceptionalism, casting hunting as anthropocentrism in practice.

In the light of Rose's perspective, hunting is nothing of the sort: this much we gather from her experience-based depiction of it. I put it this way because, surprisingly, Rose does not enter into any discussion with the anti-hunting standpoint just referred to, a standpoint she surely must have encountered on countless occasions in her academic career. The fact that Peter Singer is not mentioned in her book attests to her lack of interest in justifying her pro-hunting position to its many detractors.

On my view, Rose's experiences as a participant in hunting are philosophically important in that they challenge key assumptions on the part of those who today categorically oppose hunting, this being more and more the default position in recent years, not least following the huge impact that Peter Singer's work has enjoyed, not only academically but in the wider public. (I shall return to this in Chapter 4.)

To begin with, Rose challenges the assumption that to kill animals is tantamount to, and indeed depends upon, viewing them as objects, as passive recipients of whatever we humans choose to do to (against) them, since allegedly devoid of the intelligence, volition, and other capacities of agency that we humans alone exhibit, at least in the required advanced degree. In other words, to kill animals means to objectify and instrumentalize them, treating them as mere means to human ends and as commanding no perspective in their own right that we have to respect.

Nothing of this is borne out in Rose's report of what it is like to kill an animal by way of hunting. The alienation – estrangement – implied in the above assumptions, separating humans from animals, perceiving and treating them in a quasi-amoral, detached, disinterested manner, adopting a stance of indifference in the face of beings (somethings, not someones) of an indifferent sort and so prompting but indifference on the part of their human counterparts – everything that goes into this stance is precisely *not* what Rose experiences in her encounter with the animal she is about to kill. Instead, what she describes is a kind of human-animal intimacy and connectedness, a togetherness in life and in death, one not known, not to be had, in other types of situations where humans and animals interact. The significance of animals for humans, the kinship that exists between them and that both partake in, is precisely what is borne out and sustained in the practice of hunting. More than practically all other practices we can think of, hunting demonstrates why animals matter to humans: rather than indifference reigning here, it's all about coming to full recognition of the difference that animals make in our lives.

To suggest that the willful killing of animals is the answer to Midgley's question about why animals matter, is perhaps a bit over the top – a success when viewed as a provocation, but with little to show in terms of argument. I am not saying that it is Rose's answer, the way she asks to be read. What is beyond doubt, however, is that by bestowing such eminent visibility on the act of killing animals, making it a centrepiece of human-animal relationships, she wants to deny the fact that we humans do continue to kill animals – protests notwithstanding – the air of remoteness and abstraction that the delegation of its operation to a handful of professionals never seen, never heard – just like the animals killed, in the factory facilities never visited – nowadays amounts to. For all the ethical grounds that people may cite in their opposition to this practice, holding that killing animals is morally wrong, period, they most often do so in a situation where that killing has been subject to a thoroughgoing *Versachlichung* such that its moral import is silenced, and indifference reigns as a result. If killing animals should be looked in the eye as the morally charged phenomenon that it truly is, then hunting as described by Rose does exactly that – factory-like killing does not. It takes one-to-one encounters to do justice to what is at stake when some creatures kill others, to make them aware of what they are doing and why, experienced here as matters of life and death, as far away from a stance and a matter of indifference as you – be it the human, be it the animal – can get. Anthropologist Eduardo Kohn explains the alternative to indifference like this: "Some notion of the motivation of others is necessary for people to get by in a world inhabited by volitional beings. Our lives depend on our abilities to believe in and act on the provisional guesses we make about the motivations of other selves" (2013: 118). Hence, "it would be impossible for people in Avila [i.e., the Runa of Ecuador's Upper Amazon studied by Kohn] to hunt or to relate in any way within this ecology of selves without treating the myriad beings that inhabit the forest as the animate creatures that they are. Losing this ability would sever the Runa from this web of relations" (ibid.). Kohn employs Stanley Cavell's term *soul blindness* to describe the various debilitating forms of soul loss that "result in an inability to be aware of and relate to other soul-possessing selves in this ecology of selves". Since all selves have soul in the ecology of selves that the Runa are part of alongside the myriad of non-human beings that inhabit the forest, soul blindness is "not just a human problem; it is a cosmic one" (2013: 117).

Rose says that if Old Tim were to formulate a commandment (which he did not), it would probably be: "Thou shalt not turn thine eyes away from the deaths of animals". Considering that Tim and his people are hunter-gatherers, such a commandment may strike us as being in tension with the killing of animals he took part in. What is more, it makes for an intriguing, and rather unexpected, contrast with the position typically taken on killing in our societies today: whereas those who like Tim do kill animals urge humans not to turn their eyes away from the death of animals, those who would never take part in such killing because wholly against it do so in a situation where their eyes are indeed turned away from the animals being killed. We glimpse the oddness of the modern arrangement where there is a division of space such as to allow killing to go on a greater scale than ever (its modus operandi having been professionalized, industrialized, marketized, the killed animals being commodified and carrying a price tag determined by supply and demand), yet allowing fewer and fewer humans to have one-to-one experiences with the killing occurring, with the large majority of vegans and meat-eaters alike living at a safe and comfortable distance from the facilities.

What I have in mind, then, is the following contrast, implicit in Rose: not only is one who kills by way of hunting the one who looks the animals in the eye; he is also the one who insists that doing so – looking them in the eye – is what gives access to the moral import of the phenomenon of killing, the import being not that it is wrongful to kill them but a necessity inseparable from the coexistence on Earth of different species (here: humans and wild animals). By contrast, the one who finds such killing wrong will typically not be on the scene, will never look the animals in the eye. Instead, the moral import of the phenomenon will be decided abstractly, apart from the context with the encounter-cum-killing takes place, as purely a matter of principle: killing animals is wrong, period, and there is no need to look anybody in the eye before deciding the matter.

It is only when a certain stage in the historical trajectory of humans-animals relationships is reached that it becomes plausible, and with time even hegemonic, to settle the moral issue posed by killing animals in a perfectly principal way and in abstraction from the experiential context where that practice of hunting takes place. The premise for reaching a conclusion this way is that the issue can be settled once and for all and without any need to take the experiential dimension – that of actual encounters and the inter-species relationships they are part of, understood as an ecology of selves in Kohn's sense – into account. Indeed, people who are introduced to hunting as a meaningful and justified practice owing to their socialization into say, a hunter-gatherer culture, will be seen as "biased" on the issue, whereas people without any personal or communal experience of the matter will be seen (will see themselves) as culturally neutral and uncommitted, hence as satisfying the requirements for an "impartial" point of view.

Why do I put such emphasis on framing the issue in this way? One reason is the Shepard-inspired one that shifts in ways of looking at animals are eminently historical in nature, deeply culturally entrenched and as such passed on through the generations as *the* natural way of interpreting the world. This being so, the "view from nowhere" whereby objectivity and impartiality are held to be secured, is nothing of the sort; it is a point of view just as much a product of history and culture as the ones it prides itself on being able to dismiss on grounds of their alleged partiality. To be sure, the genesis (origin) of a view is one thing, its validity (claim to truth) is another, and they should not be conflated. Granted. But in the matter under discussion here, the contrast between an experience-based point of departure and an experience-suspending one that takes abstraction to be the one essential requirement for deciding the issue is not only particularly striking. It also impacts on the very identification of the phenomenon at hand: the nature of a person's real-life relationship with the animals we are talking about makes a big difference: concretely, it is the difference between a lifeworld where these animals are a crucial presence, and one where they are largely absent.

Against this background, we may resume Rose's account of hunting:

> Animal deaths within a kinship system are situated within relationships of accountability. In the first instance, one is accountable to the human relations of the animal, for it is they who will take direct action if their kin are being wronged, by overhunting, for example. More widely, however, every decision to kill an animal takes place within a wider set of relations signaled by the term "good country". The purpose of killing animals is to nurture humans; it is not to eradicate animals or to wreck country. Good country is a flourishing set of relationships – interdependent and mutual.
>
> (2011: 26)

Notice how the accountability of humans who hunt animals is intertwined with the capacity to "take direct action" on the part of the animals, so that the "wrongness" of overhunting is something the animals will communicate to the hunter at fault. The response forthcoming from the animals who are being wronged and who perceive themselves to be so takes precedence over the response forthcoming from the hunter's comrades; or rather, to put it more precisely, the human (communal) rebuke is such as to follow through on, and to gain its justification from, *the wronged animals*. What qualifies as morally wrong, then, is not settled, is not first and foremost played out, within the relationships of the humans involved, but within the relationship between the humans and the animals – the inter- and cross-species context is not only that of origin (where the transgression took place); it is that of its determination and its being settled as well, requiring as we see the agency exerted by the animals no less than that exerted by the humans. Everything comes to pass in the experience-based, encounter-driven inter-species context where the offence occurs. There is no resort to a sort of abstract (disengaged, neutral, external) point of view either to settle the wrongness or to respond to it in appropriate manner. The inter-relational embeddedness and rich contextuality of the *points of view* (notice the plural) in play here is the precondition for the moral issue to be settled, not a circumstance that would undermine the possibility of judgement. The many points of view in play are those inhabited by those involved, be they humans or animals; that is to say, points of view of parties that are not interchangeable but instead indispensable, and who recognize each other as precisely that.

IV

Rose observes that in hunting, "the purpose of killing animals is to nurture humans" (2011: 2011). But doesn't this go to show that, when everything is said and done about hunting exhibiting an animist cosmology, the essence it boils down to is the otherwise dismissed anthropocentric one, that of meeting a purely human need?

It is tempting to say "yes, indeed". But it would be to miss out on the peculiar dependence of humans on the animals hunted, a dependence that exposes the sort of anthropocentrism that implies human superiority to be out of step with reality – *hubris*.

I shall bring Aldo Leopold into the discussion here, author of the still hugely influential *Sand County Almanac*. Leopold is an interesting figure in that, besides being a pioneer ecophilosopher he was also, throughout his life, a dedicated hunter, and he "never suffered from any guilty conscience on that account", as the anthropologist Philippe Descola (2013: 196) remarks. Running through Leopold's reminiscences about his experience with hunting is one fundamental theme: knowing how to hunt is knowing how to find one's game and "knowing how to find it is knowing how to adopt the point of view of the animal that one is seeking, perceiving things as that animal does and putting oneself in its place". It thus involves, Descola continues – and this is his account, not Leopold's – "abandoning a superior vantage point in order to seize, from within, upon this tangled web of destinies and desires that weaves together the world in motion" (ibid.). Descola goes on to provide the wider anthropological context, observing that "such an attitude is reminiscent of the manner in which Amerindian hunting communities envisage the metamorphoses that mark relations between humans and nonhumans, such as exchanges of viewpoints in the course of which each party, modifying the observational position imposed by its body, endeavors to slip into the skin of the other in order to see things from its point of view" (ibid.).

Descola aptly warns against overstating the analogy between a twentieth-century American forestry engineer-cum-ecophilosopher going hunting and a group of Amerindians doing so. He points to the fact that Leopold's celebrated land ethic "in no sense calls into question the ontological distributions of naturalism, which it, on the contrary, accepts without a qualm" (2013: 196). The essence of Leopold's land ethic is captured in the often-cited statement that "A thing is right when it tends to preserve the integrity, stability, and beauty of the biotic community. It is wrong when it tends otherwise" (Leopold 1949: 224f.). To be sure, writes Descola, there is an educational need to "think like a mountain", as Leopold teaches, "the better to evaluate the balance between wolves, deer, and vegetation to be respected on its slopes" (ibid.), this being important to the professional forester and the hobby hunter alike. To Descola, though, it is no more and no less than "a matter of experiencing salutary thoughts of a kind to give substance and a live urgency to abstract ecological learning; in no sense does it constitute a profession of animist faith" (ibid.). Leopold, that is, never "imputes to nonhumans an interiority analogous to that of humans", for the awareness of a future with which he credits them is nothing but a metaphor for the general teleonomy of nature, which, he believes, reverberates within each one of them. Above all, he never ascribes to animals or to plants any ability to lead the existence of a species characterized by cultural conventions, since, for him, the latter are strictly the prerogative of humans (2013: 196f.). For all his celebrated "radical" ecological thinking, Leopold remained loyal to the deeper underpinnings of Western thought about culture and nature, humans and animals; he never gave up on the tenets of the naturalism informing his training as a forestry engineer.

The contrast Descola draws between Leopold's land ethic and an animist cosmology is instructive. It helps us realize that we misunderstand Rose's remark that "the purpose of killing animals is to nurture humans" if we take it to reveal some kind of closet anthropocentrism on the part of hunter-gatherers like Old Tim. Since humans hunt animals for the purpose of eating them both within an anthropocentric lifeworld and within an animist one, this fact in itself is ill-suited to let us realize the deeper differences between them. We have become so accustomed to the *moral* issue (killing or not killing) taking precedence as the truly important issue about the human-animal relationship that we tend to forget about the epistemic and ontological dimensions on which the morally "appropriate" one is premised: the ways in which humans and animals matter to each other and, as part of that, perceive each other; in particular, why it matters how animals see us, not only how we see animals. Eduardo Kohn, Descola's colleague, reminds us that "encounters with other kinds of beings force us to recognize the fact that seeing, representing, and perhaps knowing, even thinking, are not exclusively human affairs". In other words, "*how other kinds of beings see us matters.* That other kinds of beings see us changes things" (Kohn 2013: 1). For this reason, Descartes would make for an altogether inept hunter were he to act upon his contention that animals are mere "automata". In taking himself qua human as the only agent in every sense of the word – intelligence, affectivity, volition, intentionality – the animals making up the game would be but a collection of natural objects, passively there for the taking. In other words, in acting upon his philosophical views about an absolute divide between humans and animals, he would approach the animals as devoid of everything it takes to constitute a point of view in and on the world and what comes to pass in it in a given situation, denying them the capacities required to take part in give and take, call and response, in interpreting what the other's calling and responding means and adjusting their actions accordingly, as exemplified in the life-and death decision-making about whether to stay put, to attack, or to flee.

To be sure, making the point about the agency exhibited in animals by way of the sharp relief provided by Descartes' blatant denial is easy. What this contrast nevertheless provides is insight into the intimate connection between perception and action, between how we view something (or someone) and how we feel it natural and justified to treat it. Is the other a subject or an object, active or passive, capable of calling and responding or not? This is not an epistemic issue in some abstract philosophical sense, to be settled *sub specie aeternitatis*. It is not only that it matters who the participants to the relationship are, in the concrete situation. And the fact that the human's "other" is an animal does not suffice, there being so many kinds of animals, with vast differences of capacities for agency between them: a worm so different from a salmon, a salmon so different from a jaguar.

Important as these differences both within and between species are, however, their significance derives not first and foremost from their capacities (properties) as tied to species in each case, but from every individual's role in a wider web of dependencies. To be able to exist, and to persist in doing so over time, is a matter of dependence-on-others, not self-sufficiency. For humans, to kill animals for the purpose of eating them is about vulnerability, admitting being at the mercy of animals who, in their role as prey (though often themselves predators), have a choice over which humans have no control: whether or not to give themselves up to the humans who need them. For humans to refuse to admit being at the mercy of the animals, is tantamount to denying them the respect that is the very prerequisite for the animals to allow themselves to be taken. This means that there is only so much the human hunter can do – and all of it reverberates around showing the appropriate respect. The rest is up to the animals.

In stressing that how we perceive something (someone) determines how we treat it, I do not think of perception naively as a first mover. How particular animals are perceived is itself a product of cultural history (Shepard) and of the ties of connectivity that the animals are experienced as being part of, as contributing to in a life-sustaining manner, which is to say a manner that involves many different species, necessarily so. To put it crudely, either animals are experienced as so many "others" that we humans are tied to in relations of connectivity, or are they not. Either we see ourselves as in a moral relationship with animals that commits us to a conduct guided by respect, humility, and gratitude, or we regard animals as excluded from such a relationship and its peculiar commitments, turning morality into a purely human-to-human phenomenon.

One may respond by saying that, ecologically speaking, cross-species connectivity is a matter of fact, a precondition of life, and as such one that must be heeded, like it or not. But this fact can be denied, be it culturally (collectively), be it individually. In present-day society pointing to a kind of connectivity that exposes human dependence and vulnerability is apt to trigger feelings of discomfort; the very act of pointing out this connectivity is perceived as a provocation, probably more so among men than women (see Gilligan 1982). The implied lack of control is unwelcome, and defence mechanisms are called upon, including outright denial. As for the role of animals in this picture, not only do they challenge the notion of human exceptionalism in terms of capacities of agency. Our dependence on creatures who are like us in many respects, and profoundly different from us in others, is another blow, the more so the more we realize that this dependence is multidimensional and makes for a historical debt: not only have we humans in the past had a need for animals to eat for the sake of survival; as Shepard shows, animals have been crucial in helping us develop our peculiar capacities qua humans, and as we shall see by way of concrete examples below, *animals put us to the test as moral agents*, perhaps even more profoundly than do our fellow humans.

A lifeworld of connectivity as understood by Rose is one where "the well-being of one is enmeshed in the well-being of others" (2011: 27). Since there is no position outside of connection, there is no such thing as a self-interest that concerns only the self; and since one's own well-being is dependent on what happens to others, meaning other species, everyone has a stake in the well-being of others – a far cry for the model of mutual indifference or disinterest, of calculating whether or not to get involved with others or to stay outside of webs of commitment advanced in contract theory from Hobbes onwards. Dependence upon others is not a negative, not a limit on freedom and flourishing on the part of the individual, but a precondition of life that exposes the notion of self-sufficiency as a falsehood, a dangerous fantasy. "Interests are mutual", says Rose; they exist and are pursued precisely in the plural, not as the prerogative or property of the individual and thereby casting "the original situation" as one of conflict with others. And while they are not indistinguishable, Rose continues, interests are situated "within the larger dance of life which involves life *and* death, self *and* others, us and them" (ibid.).

It is noteworthy that in statements like this Rose makes no mention of the distinction between humans and animals. That is to say: the logic of connectivity and the facts about dependence on (species) others apply to both, at least on the level intended here, that of conditions of life. But whereas the facts about such connectivity as an ineluctable, non-optional given, for humans as well as for other species, can be stated categorically as Rose does here, she is also acutely aware of the extent to which the acknowledgement of this connectivity differs between cultures and from one historical period to the next. That humans and animals alike are situated "within the larger dance of life which involves life *and* death, self *and* other, us *and* them", certainly resonates with Old Tim, as it does with the Achuar in Brazilian Amazon studied by Descola and the Runa in Ecuador studied by Kohn, representing three distinct versions of an animist cosmology. But what about people living in big cities, be it in Australia, Brazil, or Ecuador? Rose speaks to this when she writes that "in contrast to the tactile immediacy of killing and eating that brought us into a region of encounter, claim, and responsibility when we were hunting, our Western contexts do not offer many opportunities to see most of the deaths on which our lives depend, either directly or indirectly" (2011: 27). Indeed, "these days hunting is the exception, not the rule, and it is so regulated by external authorities that it does not require a foundational understanding of connectivity and a capacity to be self-regulating". Having never gone hunting ourselves for lack of need or opportunity, or because opposed to hunting on moral grounds, most of us buy our food in the supermarket. In doing so, Rose points out, we are implicated in contemporary animal farming, "the mainstream methods of which involve an amplification of the monstrosity of animal deaths" (2011: 28); a monstrosity we, at least to a degree, know about, but most of us don't experience it – we don't need to, and we don't want to. In this way, "the industrialized manufacture of corpses for food is sustained at the expense of hundreds of thousands of other lives", yet most of us are not disturbed as long as this "monstrous cruelty" is hidden within organized invisibility: killing of animals no doubt takes place, in massive numbers as compared with the hunting practices depicted above, but it does so in the *modus operandi* of silent killing.

V

What happens when death is hidden from view? Is the endeavour to hide death an attempt to deny it? Or is the point of hiding it to rob death of its moral sting?

Not all death is a result of killing, though it is so in the case of hunting. Death is a fact of life, only sometimes caused by killing. Does it take killing to make death a moral issue, making it a matter of the killer's responsibility? Is it only when the one causing death by killing is a human person that it becomes suitable for moral judgement? And what difference does it make if the one killed is an animal and not another human being?

We have seen that Rose, in pondering these questions, puts great emphasis on context and circumstances. For her, humans killing animals is never morally unproblematic, nor is it ever morally neutral. Not fearful of being read as romanticizing hunting – something she must know many a reader will do – Rose clearly thinks that the killing that takes place here is both more true to the facts (the conditions of life) and more in touch with the moral issues involved insofar as the killing occurs in a context of physical encounter. By comparison, the killing of animals that goes on in industrialized fashion behind walls is part of an attempt to take away the moral sting, to empty what is being done of the moral charge intrinsic to encounters, to proceed in an atmosphere of indifference and neutrality whereby efficiency is the yardstick that counts. In his important book *Modernity and the Holocaust*, the sociologist Zygmunt Bauman coined a word for this transformation: adiaphorization.

Bauman used the term for what happens when humans are killed in a well-organized, large-scale manner, not animals. Whereas he was addressing genocide (the Holocaust), Rose is addressing extinction. Although many creatures are affected by extinction, she is mainly interested in animals, so much so as to define her take on the Anthropocene, that "new era of solitude", largely self-inflicted at that: "As Earth others depart, never to return, we face a diminishing and impoverished world, and equally, we face new, agonizingly, lonely, questions about the meaning of our existence" (2011: 10). This means that for Rose, the imperative of our time is not what it was for thinkers like Bauman, and Adorno and Jonas before him, namely "never again Auschwitz". Instead it is "to recover or discover connectivity". That can only be done by recovering "the radical awareness of being at home that emerges as we embed ourselves ever more complexly into the life of the world". And she adds, "the world is suffering, life is dying, and the project of embedding ourselves exposes us consciously to peril" (ibid.).

There is an urgency to this imperative that is different from the Holocaust-related one. The urgency is to do with the imperative being not only (exclusively) ethical, but its being *ecological too*, the one inseparable from the other, whereby any wish, or attempt, to separate the two would mean to get it wrong about the situation in which ethics is – from now on, if not before – embedded.

Rose clearly views the Anthropocene as the age of extinction. True enough, some entities expand and see growing numbers – humans and human-made artefacts. The problem is that the expansion of one species comes with the cost of decimating so many others. Rose puts it like this: "In a few short centuries, the human species has begun unmaking the balance on earth between life and death, enabling death to expand and expand, tilting life toward a catastrophe that is difficult to imagine, difficult to think, and yet morally imperative to consider" (2011: 61).

It is as if death is at the same time everywhere and nowhere. It is all over the place in that species and life forms in record numbers disappear to the point of going extinct; instead of life being replenished, it is replaced by death. On the other hand, the ways in which death prevails over life, absence over presence, are for the most part invisible. Yes, animals disappear, that's what the statistics tell but we don't actually see them doing so. They are casualties of the ever-growing impact that humankind currently exerts everywhere on Earth. But for the most part their actual *dying* is without witnesses, in stark contrast to animals being killed in hunting or by way of other sorts of encounters with the human agents who cause their death, deliberately at that. And insofar as these animals die without encounter, their deaths lose the quality of being an event, a change occurring in the world at some particular place and point in time and so marking the shift from a presence to an absence.

But doesn't putting it like this smack of anthropocentrism? Well, it is for us humans that the animal deaths we are speaking about lose their quality of being an event, not for the animals themselves. Whether or not humans are on the scene when the last eagle owl departs from its last habitat along Helgelandskysten in Norway hardly makes a difference to the owl's experience: the change from this species' presence to its absence in the area is an objective fact regardless of it being noticed subjectively by humans or not. This may sound trivial, but it is not. For the fact is that the departure of animals to the point of their disappearance altogether from places they have used to live for centuries or even longer, would not go unnoticed in previous times like it does now. It takes familiarity with the animals in question to notice that their numbers are dwindling, that in places where one could expect to see them, one no longer does, and so feels lucky for every sighting, adjusting subjectively to the change happening: from a situation where one took spotting them for granted to one where doing so has become spectacular, a rare and true gift, as such to be shared with others (meaning on social media these days).

So yes, we feel thankful when spotting an animal now increasingly rare. But wait a minute – should we not be enraged instead, enraged because of the loss entailed in their ongoing disappearance? Are we lowering our expectations, learning to live with disappointment? If so, at what cost?

Since when could one no longer take their presence for granted? How to catch that moment, marking the change? But it's not a moment we are talking about, precisely not. Instead it's something evasive, part of processes originating and set in motion elsewhere, far away, one cannot even be sure exactly where, when, and by whom, to the extent that particular agents can be identified at all. The causes for the disappearance of the eagle owl from one particular area are diverse, spanning the entire continuum from the abstract to the concrete. When I read in the papers that this kind of owl is now being put on the list of "critically endangered species" in Norway, no specific causes are mentioned save the general and nowadays ubiquitous one of "climate change". This is abstract in the sense I am getting at here, despite its being true. It is abstract in that the questions about the "where, when and by whom", about the causes as *particular* in time and place and in terms of agency, are left hanging in the air. At the level of generality, of being operative "everywhere" and "all of the time", climate change is an example of what Timothy Morton (2013) calls "hyper-objects": something all-too-real in their effects yet whose exact origins – spatio-temporally and agentially – elude us, not for being few or sparse, but for existing in excess, making it difficult to identify anything, anywhere, at any time, that would today be *beyond* their reach, making the particular instance of disappearance *overdetermined.* In the Anthropocene, the privilege of something or someone existing *outside* is lost in reality, although it may live on in/as a fantasy, feeding high-tech, Elon Musk-like efforts to flee the Earth and retreat to Mars or some other hoped-for planetary refuge.

VI

There is a more sinister aspect to the disappearance of dingoes discussed in Rose's book, one that merits discussion.

First, we need to understand more precisely what kind of animals dingoes are. They arrived in Australia from Southeast Asia some 5,000 years ago. Neither clean-cut wild animals nor ordinary pets, the dingoes have provided companionship of a unique sort for Aboriginal people, being the first non-humans to answer back. However, their ambiguity in not quite fitting into either of the wild/domestic/pets categories has caused some Aboriginal people to be situated ambivalently between loving and loathing them; some want to keep them as pets, others to kill them, considering them as vermin. Rose mentions the effort of new settlers to eradicate dingoes, based on viewing them as detrimental to sheep and cattle and thus a nuisance to be removed. In contrast, and as we would expect, Old Tim was completely against not only the willful killing of dingoes, but any effort to "control" them. Tim is able to understand their language, and he can convey their points of view to others. It is part of his overall outlook that we humans come from the dingoes (and other animals), that they live on after their death and that today, owing to the effort to drive them away from the land, ancestral dingoes give voice to their sense of lost reciprocity, saddened by the betrayal of humans.

In modern times, dingo-proof fences have been erected to keep dingoes away from areas were local populations had been annihilated. Starting in 1947, state governments have built a fence which is now 5,400 kilometres long. In itself, the fence does not eradicate the animals; it is designed to control their movements (see Netz 2004). Widespread practices today are shooting and poison. The use of 1080 poison (sodium monofluoracetate) that has become commonplace is not only lethal to the dingoes, but to other animals as well; any animal that eats the remains of an animal that has been poisoned is at risk; there is no known antidote. The method used is to drop by plane chunks of dried meat that are laced with 1080. In addition, land clearing causes dingoes to turn towards sheep and calves. As a consequence of the various methods employed, remaining dingoes now "live in a state of perpetual social disorder" (Rose 2011: 69); with their habitats destroyed and their regular prey eradicated, they are forced to turn to domestic animals.

It makes no difference to the pastoralists who are waging war against the dingoes that research on their ecological impact shows that dingo populations work to "sustain a balance of species that is viable in the long run"; that there is "a strong correlation between healthy dingo populations and biodiversity as indicated by the flourishing presence of native species" (Rose 2011: 67). For example, the dingoes are crucial in controlling foxes and cats. In short, the more the dingoes are given the chance to thrive in their natural habitat, the greater the biodiversity. Rose quotes the following statement by Adam O'Neill, based on a lifetime of experience with the dingoes in Australia: "I believe that the dingo is our only chance for eco-reconciliation" (Rose 2011: 68).

How, then, to explain the killing of dingoes that goes on and on? Why don't the many benefits their presence brings carry any weight for the people who pursue their disappearance? Are there important lessons to be learned here, pointing beyond the particulars of this place, these animals, and the people involved?

I certainly think so, but before we get to the bigger picture, where the ethics of Emmanuel Levinas will take centre stage, we must explore in more detail what Rose is reporting.

We have looked at death by poison, but we have not discussed the method where a dingo has been shot and hung over the fence, for all passers-by to see. Upon realizing that "the dead body had been casually but carefully draped across the fence", Rose asks: "Was it a message, and if so, to whom was it addressed?" (2011: 92). She passes the question onto a friend who knows the pastoralists better then she does, getting the following answer: "He does it because he can". Notice that they both – two women – take it for granted that the person responsible is a man. As far as I can tell, they are justified in doing so. Nevertheless, it leaves something unsaid: could a woman have done it? If not, why?

Rose's further reflections are worth quoting:

> In life, these creatures were despised by the pastoralist, and now the death is displayed in a narrative of total power. The pastoralist wants to destroy the animals, and he wants to display his dominance. This dingo, this outlaw, can be defiled with relative impunity. Social power as well as interspecies power is on display here. This exhibit offers up trophies in

> the war against the wild and living world. (…) The bodies are trophies, and thus are mirrors: they give us ourselves. They give us ourselves as we are when we go out to torture and kill, and are proud of our work. […] Dingo trees amplify suffering further. The trophy display seems to require the victims to state that the harm has been a positive good in the world, that somehow their suffering and death are something for the killers to have been proud of. […] Trophies form a "narcissistic mirror" in which the killers can read their own moral superiority, for it is they who have made the world a better place.
>
> (2011: 93, 97, 98)

When I taught Rose's book as part of my ethics course for master students of philosophy, I showed the class the pictures of the hung-up dingoes. I also displayed the cover photo of a dingo Rose encountered in outback Queensland, described by her as "so young and naïve as to be almost heartbreaking" (2011: 68).

I remember the atmosphere in the classroom when I put the book down on my desk, not offering any comments on my part and asking if somebody wanted to say something. Silence.

How to make the connection between the dingo hung-up and killed, and the dingo "so young and naïve as to be almost heartbreaking"? You may go back and forth, back and forth, visually and in thought, between the two pictures, realizing that – in principle of not in practice in this particular case – it could well be the same individual dingo in both pictures, corresponding to life as the "before" and death as the "after", mirroring the sequence to which the killer subjected his victim. Call it heartbreaking, call it unbearable, call it a mystery: how is it possible to willfully kill this animal, to respond by inflicting death when faced with such an expression of openness, trust, affirmation of life, a welcome to come closer, come in, share, and reciprocate?

Rose doesn't put the question like this, but she might well have done so, not least because she chooses to consult the ethics of Levinas – highlighting as it does the peculiar authority of the face, of the appeal emanating from it, a wordless "do not kill!" – to provide a framework for her further reflections. To anticipate a point I will elaborate in later sections, consider here the following suggestion: the constellation I just described does not amount to the flagrant contradiction we intuitively take it be. What does that mean? It means that the dingo is not killed *in spite of* the expression of openness and trust with which it greets the human approaching, be it with a camera in hand, or a gun; the dingo is killed *because* of it.

Probably many will respond by saying that, if anything, *this* makes for a flagrant contradiction. But as we will see, evildoing often is just that: the chosen response to goodness. Again, this is my suggestion, not Rose's. But I see it as making her own point, only put differently – hers being that what we have to do with is *death as the response to life*.

To appreciate what is at stake here, we need to return to Rose's question to her friend: what is the person saying in putting the body of the killed dingo on display for all to see?

On my view, the person doing this is conveying a message along several dimensions. First, there is the message to the individual dingo in question: I am free to kill you, and having killed you I can do with you what I like. I can do so with impunity, demonstrating that I am sovereign in my relationship with you – so great is my power that I don't accept you as my counterpart at all. In killing you this way, I act upon my belief that you are not worthy of being in any relationship with me at all. I can do without you; the world is a better place without you. Having the right to do this to you means that I have the right to do the same to all of you, your whole kind; there is no difference between you and other dingoes. All dingoes who see you hung up like this will know about the death that awaits them; they will know what to expect from humans. Second, there is the message to members of his own kin, his own species, be they fellow pastoralists, be they outsiders like Rose, that this is what people like him are entitled to do whenever encountering an animal like the dingo. In deciding to put the bodies up like trophies, the message is sent that rather than this act of killing being something to hide or be ashamed of, it should be looked upon as an act of goodness on behalf of the community – protecting humans from animals at the wrong place doing the wrong things – and so as matter of rightly felt pride, apt to put on display. Third, in demonstrating the refusal to recognize the animal, recognition among fellow humans is sought and – in all likelihood – found: its demonstrated denial to non-humans consolidates its status as a humans-only affair. As far as dingoes are concerned, excluding them from the moral universe is a means to the end of sustaining a purely human-centred moral and social order. In seeing the bodies of dingoes hung up on trees or over fences, everybody – humans, animals – can consider as confirmed that the world is set aright, that things are as they should be.

This raises questions about the relationship between life and death, increasingly precarious and out of joint; questions about extinction and about whose lives are grievable that I will investigate in depth in later chapters. Before concluding the discussion of Rose's book, however, we must examine her commentary on Levinas' ethics.

VII

Emmanuel Levinas is the philosopher Rose turns to in her search for an ethics that does justice to the relationship between humans and animals like the dingoes, helping us to identify what has gone wrong in a case like the one Rose witnessed.

As it happens, there is an essay by Levinas that seems ideal for the task, entitled "Name of a Dog". Born in 1906 and growing up as part of the vibrant Jewish community in Vilnius in Lithuania, Levinas took up residence in

Paris, enlisted in the French army, was captured and put to work in a forestry commando unit for Jewish prisoners of war in Nazi Germany. The work group Levinas was part of was for a brief period "adopted by a dog" (this is Rose's expression; 2011: 29); they named him Bobby, and he would enthusiastically greet the men in the morning and bark excitedly as they returned from work in the evening.

As we would expect, Levinas writes lovingly of Bobby when he revisits his memories of him in the essay written 30 years later. But fond memories do not prevent Levinas from stating that "dogs are without ethics and without logos", proceeding to proclaim that Bobby was "without the brain needed to universalize maxims and drives" (Rose 2011: 29f.; see Levinas 1990: 152, 153).

What are we to make of these statements? Rose is shocked. Levinas' ethics being famously founded on the authority of the face, issuing the commandment "thou shall not kill!" and so placing the moral subject in a non-optional, exit-forbidding position of absolute responsibility for whomever is encountered as a face, Rose expected that Bobby would be acknowledged as precisely such a face with such a moral authority to it. And yet this is the recognition that Levinas seems intent on denying Bobby. Why? Because Bobby is a dog, not a human person. In being "without the brain needed to universalize maxims and drives", Bobby fails to qualify as a face in Levinas' sense; to be different from humans in regard to these capacities is tantamount to exclusion from the moral universe. Face means the face of a human; no other face, the face of no other, will do.

To Rose, "Levinas' rejection of Bobby is anguishing because the greatest twentieth-century philosopher of ethical alterity could not unambiguously make a place in his ethics for the one living being who approached him and others with full recognition of their humanity" (2011: 30). In denying Bobby a face in the sense on which everything ethical depends, and from which everything to do with being called to respond by way of responsibility flows according to Levinas' teaching, he effectively "restores the tyranny of the abstract over the living reality of the world" (ibid.), reinstalling logos as what differentiates humans from animals, leaving Bobby powerless to gain ethical entry, no matter how lovingly and trustfully he greets the humans he gets to know. Evidently disallowing any dialogue, instead insisting on the radical and unchangeable non-symmetry and non-reciprocity between himself and the dog, Levinas leaves his readers with a frightening reminder of the "close parallels between animals who are killed with impunity and humans who are killed with impunity" (Rose 2011: 31). How can it be that Levinas, who dedicated his ethics to the memory of the six million Jews who perished in the Holocaust, offers what is in essence a "replication of the structure that underlies the possibility of genocide"?, asks Rose, making no secret of the depth of her disappointment (ibid.).

Rose ends on this note, citing Luce Irigaray's statement about the "devastating emptiness" to be found in Levinas' work, said to hinge on his dedication to the abstract, demonstrated in his assertion that "the best way of encountering the Other is not even to notice the color of his eyes" (2011: 39). When everything is said and done, and contrary to expectations raised by foregrounding the ethical significance of "alterity" and "exteriority", of the ethical Other subverting the omnipresent logic of "the same", Levinas effaces difference. That Bobby fails to qualify ethically, as a face in Levinas' sense, simply for being a dog rather than a human, proves the point beyond doubt.

I shall not engage with the many critical commentaries that Levinas' essay about Bobby has provoked, some of which Rose draws on. Like the commentators, she directs attention to the places in his text that appear ambiguous – why, for example, does Levinas himself point to the impunity that holds for both killing humans and animals? More to the point, why isn't he taken aback by the parallel? Why doesn't he allow it to show up his exclusion of Bobby from the moral universe as just as dangerous, just as morally wrong, in principle, as the exclusion of the Jews by the Nazis? Why allow, let alone advocate, difference to work *against* an entire group of beings, be they animals, be they humans?

These are my questions, not Rose's, although I take them to be very much in her spirit. However, there is one instance of ambiguity that I don't see Rose paying sufficient attention to, one I consider crucial to coming to grips with what Levinas is up to in his essay.

Rose tells us that "Levinas calls Bobby the last Kantian in Nazi Germany" (2011: 29). Why would he do that?

We need to know: what does a Kantian do – here meaning: what is it that Bobby does that makes him a Kantian? And what makes him the last one? My answer – which I find neither in Rose nor in Levinas' text – is that Bobby shows himself to be a Kantian when he does to them what the Nazis make a point of not doing: he treats them as ends in themselves, thereby recognizing their inherent worth, the *Würde* that Kant considers the hallmark of their inviolability in moral terms. As opposed to the relativity and commensurability of an entity's *Wert*, allowing for trade-offs and compromises, a being's *Würde* speaks to its unconditional, categorical, and irreducible worth qua what it is in and of itself. This is Kant. It is what he says about humans, not animals. But it is Kant nonetheless.

In at least one respect, then, Levinas acknowledged that Bobby was a Kantian – namely, in the way he greeted (treated) humans such as Levinas and his workmates, in stark contrast to their being denied being treated that way by the Nazis. Evidently, then, Bobby didn't need to have "the brain needed to universalize maxims and drives" in order to actively recognize the individuals comprising the group of prisoners – recognize them the way there are morally entitled to qua ends in themselves, at that. Here, Bobby passes the test while the Nazis fail it. The dog does better, morally speaking, than this group of humans. If somebody in this inter- and cross-species constellation has anything of merit to teach the others by way of example, it is the dog. Bobby walks the talk without even having to talk.

When Rose quotes Levinas as stating that "dogs are without ethics and without logos" (Rose 2011: 28), I am not quite sure how to interpret his meaning. Rose offers no explanation, nor do I see the commentators she draws on doing so. Let me suggest, for the sake of argument, that we grant Levinas the claim he makes about dogs not having the brain needed to universalize maxims and drives, and that Bobby, in that sense, is without logos. The question is: so what? Pressing on, we need to ask: are dogs "without ethics" *because* they are without logos?

If "because" is what Levinas means, why doesn't he say so? Why does he say "and" instead? And if "because" is, nonetheless, taken to be what he means to say, does it qualify as a valid proposition? Does the "without ethics" follow, compellingly, from the "without logos"? Well, only if the possession of logos is a necessary condition for the possession of ethics, for being what Alice Crary terms "inside ethics" (2015). I see no argument being made to this effect in Levinas' text.

It is easier to understand what is meant by being without logos than what is meant by being without ethics. The latter is both an unusual and an inexact formulation. One interpretation would be that Bobby is without ethics insofar as, and thus because, the humans who encounter him treat him as a someone not worthy of moral consideration. Though this denial of recognition says a lot about these humans, it doesn't say anything about Bobby. He comes up short for reasons concocted by others, members of a different species, reasons that are arbitrary and selective in focusing only on what he is deemed to be lacking, ignoring what (who) he positively, demonstrably, is, what he is capable of doing – for instance, greeting a group of humans deliberately denied greeting by another. If anything appears odd here, it is the thinking and the practice of the humans involved, not that exhibited by Bobby.

Let me arrive at a conclusion. Bobby may be without logos (in Levinas' philosophical sense, going back to Plato), but that lack did not prevent him from treating Levinas and his comrades in a way worthy of the last Kantian in Nazi Germany. If there is criticism involved here, its proper target is certainly not Bobby, but the Nazis. Bobby was able to do what is morally expected, even mandatory, for anybody who encounters men like Levinas: to respect them as beings qualifying as ends in themselves, being no mere means to others, be it Bobby himself. He could pull this off without possessing logos, without the brain needed to universalize maxims and drives; indeed his chosen action did not have to pass some such (cognitively) sophisticated procedure or test, it was much more simple, elementary even, no need to equivocate over whether to greet or not. Clearly, then, Bobby's doing what he did must count as partaking within ethics, not as being "without ethics", somehow outside it: Bobby does his part, even though the humans (or some of them) fail to do theirs.

VIII

In holding that what Levinas says about Bobby represents a "moment of truth" about his proposed "ethics as first philosophy", devastatingly exposing its blind spots and limits as regards the status of animals, Rose is far from alone. Luce Irigaray's harsh criticism was mentioned, contributing to what I understand is now emerging as the "majority reading" (Rose's term) of Levinas' position on animals.

To be precise, Irigaray's critical focus is not on animals. When posing the question "Who is the Other?" to Levinasian ethics, the specification she adds is "if the other of sexual difference is not recognized or known"; it is not "if the other is an animal [e.g., a dog] instead of a human being". The crux of Irigaray's critique, then, to quote a formulation of hers not cited by Rose, is that "Levinas abandons the feminine other, leaves her to sink, in particular into the darkness of a pseudoanimality, in order to return to his responsibilities in the world of men-amongst-themselves. [...] The feminine other is left without her own specific face. On this point, his philosophy falls radically short of ethics" (Irigaray 1991: 112, 113).

Note that to drive home how Levinas "abandons the feminine other", Irigaray charges him with leaving her to sink "into the darkness of a pseudoanimality". In using this expression, Irigaray alludes to the association between women and nature that is a long-standing characteristic of Western philosophy, routinely opposed to that between men and culture (see Lloyd 1993), and subject to thoroughgoing critique in Simone de Beauvoir's groundbreaking *The Second Sex* (1949), recast there as the dualism between men as exhibiting transcendence, women as exhibiting immanence – and since immanence means being confined to the reproduction of life (giving birth), women's alleged difference from men is partly (not exclusively) to do with their similarity with animals: the cycle of reproduction points to the way in which humans are beings of nature, not beings of culture (see Arendt 1958).

Clearly, then, the question "Who is the Other?" is a complicated one for Levinas' ethics, inviting critical scrutiny in terms of gender as well as the human/animal dichotomy. Having pointed to the tendency to associate women with nature and so (indirectly or directly) with animals, I now turn to a more detailed assessment of Levinas' position on animals. I do so with an ambition to find in Levinas an insight I consider overlooked in both Rose's and Irigaray's criticisms.

For what it's worth, I spent a fair portion of the 1990s reading Levinas. I even went to Paris to do an interview with him in his apartment, in 1993, two years before he passed away (see Vetlesen & Jodalen 1997: 53ff.). It says a lot about my interest in his ethics that, even though "the face" and "responsibility" were the central topics, it didn't occur to me to raise the "animal question": to ask Levinas whether he holds that an animal may be "a face" and so a *bona fide* ethical Other. The point I want to make in mentioning this is a general one. My lack of interest in the animal question back then is reflected in virtually everything written by commentators – be it approvingly, be it critically – on this thinking in the 1980s and 1990s, the latter being the decade his ethics shot to academic fame and became an intellectual fashion, in no small part fuelled by the attention Jacques Derrida was paying to it. For example, in the widely read conversations that Philippe Nemo had with Levinas, published as *Ethics and Infinity* in 1985, there is not a single question that mentions animals. Even more noteworthy, this also goes for *The Cambridge Companion to Levinas*, edited by Simon Critchley and Robert Bernasconi and published in 2002, 12 years after his essay about Bobby first appeared in English.

Given this "absent animal" background, what is the insight I believe can be found in Levinas' thinking, and that I consider important precisely with regard to animals?

Let me I quote from the aforementioned conversation with Nemo, where Levinas states:

> I understand responsibility as responsibility for the Other, thus a responsibility for what is not my deed, or for what does not even matter to me; or which precisely does matter to me, is met by me as face. [...] Since the Other looks at me, I am responsible for him, without even having *taken on* responsibilities in his regard. [...] The proximity of the Other...is a structure that in nowise resembles the intentional relation which in knowledge attaches to the object – to no matter what object, be it a human object. Proximity does not revert to this intentionality; in particular does not revert to the fact that the Other is known to me. [...] The face orders and ordains me. The face is exposed, menaced, as if inviting us to an act of violence. At the same time, the face is what forbids us to kill. The relation to the face is straightaway ethical. The face is what one cannot kill, or at least it is that whose *meaning* consists in saying: "thou shalt not kill!". [...] The intersubjective relation is a non-symmetrical relation. In this sense, I am responsible for the Other without waiting for reciprocity, were I to die for it. Reciprocity is *his* affair. It is precisely insofar as the relationship between the Other and me is not reciprocal that I am subjection to the Other; and I am "subject" essentially in this sense. [...] I am I in the sole measure that I am responsible, a non-interchangeable I. I can substitute myself for everyone, but no one can substitute himself for me.
>
> (Levinas 1985: 95, 96, 97, 86, 87, 98, 101)

It would require a book, or at least chapter, in its own right to investigate all the claims Levinas is making in this dense passage. For my present purposes, what I consider most important is this: *One does not have to be a moral agent in order to be a moral addressee.* I take this to be the thesis that follows from the two main assertions we just saw him make: first, that responsibility is imposed on the I by (from) the Other, it is received before being thought, intended, willed, before being considered at all. And second, that the responsibility thus received from the other is unidirectional: it is *for* the other, directed at the Other without any regard for the Other's reciprocating it or not. Being prior to cognition, to the seeking of knowledge, as well as to volition and intentionality, responsibility is also prior to freedom – or, more precisely, it is what institutes freedom by giving it a task, a direction, something (someone) to be free for – the Other is freedom's *raison d'etre*, not what limits, undermines, or frustrates it (Hobbes, Sartre). In the proximity of the face-to-face encounter, responsibility becomes individuated, concerned as it is not with any other, or some anonymous other(s), but *that Other* over there. True, responsibility is absolute, unconditional, and so universal, applicable to all. Yet, *in concreto*, and this is where we see Levinas' signature, responsibility comes into being as *mine*: being non-transferable, responsibility helps constitute my uniqueness, rendering moot the very notion that someone else could step in for me as well as the concern about what others would do in my place. There is only this one place here that matters, the place that I occupy and that is my springboard for directing my attention, orienting my action: to the Other addressing me.

How does this exposition of what I take Levinas to mean by his central concepts – responsibility and "the face" – help us assess Rose's conclusion that his ethics represents a "rejection of Bobby", that it denies the animal(s) a face and so fails spectacularly when put to the test that the animal question subjects it to?

I shall nor repeat my above argument to the effect that in calling Bobby "the last Kantian in Nazi Germany", Levinas is paying Bobby a compliment. But what about that other statement, so decisive to Rose's harsh criticism, that Bobby was "without the brain needed to universalize maxims and drives" (2011: 30). First, let me repeat my "so what?" Second, I suggest that Bobby, as portrayed for us by Levinas, was perfectly able to treat his (human) others with the recognition that matches their ethical standing – so evidently, based on Bobby's performance, this can be done without the sort of brain that engages in operations like universalization. To the extent that a philosopher like Kant considers the possession of such a brain a necessary condition for recognition, Bobby's example shows that he is wrong. Moreover, as we just saw, Levinas locates ethical subjectivity in something *prior* to engaging in thoughts like "if I were to act in that way, could my action be universalized, and would others reasonably assent to it?" To do without that sort of cognitive deliberations *in foro interno* on the part of the subject is precisely what characterizes the Levinasian, as distinct from the Kantian, ethical subject. Again, Bobby is *within* Levinasian ethics, not without it or to be excluded from it. Last but not least, since Levinas is known to mark out his own position in ethics by way of distancing himself from Kant (and numerous other representatives of the Western canon), why on earth should he suddenly, when discussing an animal such as Bobby, change sides as it were and endorse the application of Kantian criteria of ethical subjectivity?

In 1992, before the debate over Levinas' view of animals got started, Simon Critchley, one of his most prominent interpreters, stated: "As Levinas writes, it was a dog, Bobby [...] who alone recognized Levinas and his fellow Jewish prisoners of war as human beings during their time in the camps in Germany. The ambiguity of ethical selfhood is that it is only as an animal, as a being that eats and is capable of giving its food to the Other, that I can be for the Other" (1992: 181). Critchley's statement is based on a passage in Levinas' *Otherwise than Being or Beyond Essence* where he writes that "the incarnate ego, the ego of flesh and blood, can lose its signification, affirming itself as an animal in its conatus and its joy"; interestingly, the example Levinas gives is that of a dog (Levinas 1991b: 81). Critchley could also have cited Levinas' contention that "Only a subject that eats can be for-the-other, or can signify. Signification, the one-for-the-other, has meaning only among beings of flesh and blood" (1991b: 74). Animals, not just humans, are such beings.

If the points I have made are granted, what do they help to illuminate? Well, they allow us to appreciate Levinas' understanding of what it means, and requires, to be an ethical subject, that is to say, the one among "the ethical party of two" (Bauman 1993) situated as the party on whom responsibility is imposed, coming from without as it were and pre-cognitively, pre-intentionally, and unasked-for at that (as explained above). In other words, what we have been looking at is whether Bobby can be said to be a subject in Levinas' sense, and the answer is yes.

In a way I would expect to be crucial to Rose, however, the discussion has been an exercise in the wrong direction, and so beside the point. Why? Because the thrust of her own interrogation is to do what that *other* party, that is to say, the question of who is to count as moral *addressee*, as distinct from subject (agent). The shock she tells us about stems from what Levinas, on her reading (shared by so many others), says about Bobby – and so, writ large, animals in general – as a moral addressee. It is in *that* respect that Rose finds Levinas' position so utterly disappointing.

As I stated above, in my view one of the strengths of Levinas' ethics is that it advocates the thesis that one does not have to be a moral agent (subject) to be a moral addressee. What does that mean? It means that the subject is responsible for the other (addressee) regardless of reciprocity; responsibility is prior to whether the other "pays back" in some form or other or not. This makes for the non-symmetry mentioned above, being of a sort that is truly radical: the weaker – more destitute, naked, alone – the Other, the greater the subject's responsibility. If anything, it is precisely the Other's *lack* of power in the form of resourcefulness that lends authority – peculiar moral authority – to the call directed at the subject: thou shalt not kill – knowing full well, both of them, that there is nothing to prevent the agent from doing exactly that, *nothing else* than the said authority.

Where does this leave us? I have argued that Bobby is to be regarded as a moral subject in accordance with Levinas' ethics – not, I need to emphasize, in the *full* sense of being such subject, insofar as that requires capacities for responsibility (and hence for such things as blameworthiness) that Bobby lacks, but in the weaker sense that is so wonderfully at display in Levinas' portrait, showing how a dog such as Bobby is able to act (yes, act) in such a way as make a difference for others, in this case the group of prisoners Levinas was part of. That being said, I have also argued that, to Levinas, one does not have to be a moral subject to be a moral addressee. So, if Bobby's being a subject (as just qualified) is not what grounds his status as an addressee, what is?

It's his vulnerability, the vulnerability he shares with humans such as the Jews and the Nazis whose company he keeps, making no moral distinction – discrimination – between them, although perhaps he would, should the one group start treating him so badly as the Nazis treat the Jews. But discrimination among the humans is not the issue here. "Only a subject that eats", we saw Levinas asserting, "can be for-the-other"; to be so "has meaning only among beings of flesh and blood". To which I should add, as does Levinas: "Sensibility can be a vulnerability, an exposedness to the other or a saying only because it is an enjoyment" (1991b: 74). Nobody can seriously dispute that such sensibility, such enjoyment, is just as real in the dog Bobby, and in other animals, as in humans. That Bobby cannot reciprocate, or will not reciprocate, the action that follows from a subject (agent) who acts out of responsibility for him and his plight, has no bearing on the ethical issue we are discussing; what he cannot or will not do is precisely what we have seen Levinas hold to be the salient defining mark of moral addresses in general. Thus there is no Levinas-internal reason to include as such addressees only humans, and to exclude animals such as Bobby. That reciprocity is uncalled for and that responsibility is a one-way affair is Levinas' central thesis, the signature of his ethics, and the reason he rightly regards it as strongly opposed to its philosophical rivals, be it Kantian deontology, Bentham-inspired utilitarianism, or contract theory from Locke to Rawls.

3 A dog named Bobby

Animals as agents and addressees

I

In *Justification and Application*, Jürgen Habermas takes up the challenge that ecological ethics poses for the Kant-inspired position he defends in his proposed "discourse ethics". His starting point is that, owing to their "anthropocentric profile", moral theories of the Kantian type seem to be "blind to questions of the moral responsibility of human beings for their nonhuman environment" (1993: 105). More specifically, this blindness can be traced to the assumption that "moral problems arise only within the circle of subjects capable of speech and action" (ibid.); they arise, that is, only between beings who share the same capacities for speech and action – and only humans are held to be such beings. The question is: Should our moral feelings, judgement, and actions be directed exclusively to subjects capable of speech and action? Or should they, in a morally compelling way, be directed to animals as well? Is there a moral obligation to protect animals?

I bracket for now the distinction that Habermas makes between humans as beings capable of speech and action, on the one hand, and animals as lacking these capacities, on the other. We simply note that this is the distinction he is committed to, as is his colleague Günther Patzig, whose work Habermas discusses in order to clarify his own position. Patzig is credited with articulating the moral intuitions that "speak an unambiguous language" in this matter – intuitions to the effect that "the avoidance of cruelty toward all creatures capable of suffering is a moral duty" and that animals confront us as "vulnerable creatures whose physical integrity we must protect *for its own sake*" (ibid.). *Contra* Kant, this means that we have duties *towards* animals, not merely in relation to them for reasons stemming from our duties to ourselves qua moral agents.

So much for the intuitions highlighted by Patzig. For Habermas, the question is: How to square them with an anthropocentric approach in which "duties in the strict sense can only follow from rules that rational beings impose upon themselves through insight"? (ibid.). According to the model upheld in Habermas' discourse ethics, only those norms qualify as valid that get, and truly deserve, the recognition of all affected parties. However,

DOI: 10.4324/9781003317692-4

as Patzig points out, since animals are not creatures of a kind that "could enter into a relation of principled reciprocity with us", what becomes the status of duties towards animals? Patzig declares himself ready to dispense with the symmetry between duties and rights that is conventionally (following Kant) wed to the reciprocal recognition between free and equal subjects, meaning human such. Accordingly, he (re)defines duties towards animals in an asymmetrical manner: "Animals do not have duties toward humans, but humans have duties toward animals" (1993: 107); hence, the greater the sensitivity to pain of animals, the more onerous duties should be. Habermas observes that this would amount to a transformation of the concept of duty as coined by Kant; "it is no longer a question of duties in the deontological sense but of the relative preferability of goods" (ibid.).

In his attempt to do justice to the intuitions he lists – crucially, that we should not neglect animals callously, let alone torment them – Patzig tries to connect the injunction to protect animals, inspired by Bentham's "can they suffer?", to the deontological theory. Patzig's argument goes like this: "It would be irrational to make a radical distinction between human and non-human creatures where the latter behave in such a way that we are compelled to assume that they are capable of experiencing pain and suffering. *Thus* the prohibition on arbitrarily inflicting suffering and callous neglect is extended beyond the human domain to include non-human creatures" (1993: 108). Habermas' comment on this statement is worth quoting in full:

> Regrettably, the "thus" here masks a non-sequitur, for the principle of generalization does not entail that the norm of refraining from inflicting pain on others should be extended to encompass *all* creatures sensitive to pain, so long as this principle demands only equal consideration of the interest of all those we may assume are capable of adopting the perspective of all others. Apart from their potential "yes" or "no" responses it cannot be determined whether such a norm could meet with the agreement of all and hence is valid. Animals, as Patzig correctly emphasizes, simply do not belong to the parties to the contractual reciprocity that underlies human morality.
>
> (1993: 108)

This is a remarkable passage. In essence, it says that Habermas sees no way in which the principle of generalization would allow, let alone demand, the norm of refraining from inflicting pain on others to encompass all creatures sensitive to pain. Why? Because the principle demands that only (exclusively) those we may assume are capable of adopting the perspective of all others be given consideration in the form of having their interests protected and respected. Indeed, the principle of generalization as Habermas understands it demands this – it demands, as he himself points out so as to refute Patzig's argument, that the moral domain be limited to "those we may assume are capable of adopting the perspective of all others". But to insist that the principle entails this limitation with regard to who should be included within the domain of those worthy of moral consideration, is to beg the question. It simply presupposes that this principle, understood this way, is the appropriate and valid principle in the matter. No alternatives are considered.

If anything, what Habermas shows is the burden of proof that comes with the decision to take *this* principle, and the requirements for its proper generalization, for granted. Note what is happening here: Patzig gives his argument for refining duties towards animals in an asymmetrical manner. Habermas dismisses the argument by pointing out that it cannot be squared with the requirements of a moral theory where ascription of moral status depends on symmetry in the form of reciprocity between agents and addressees. This is precisely the type of theory – the position – Patzig is out to challenge, wanting to show up its blindness towards animals. How can Habermas reject the case for an asymmetric, non-reciprocity-requiring moral theory by saying that it does not conform to the symmetric and reciprocity-requiring one?

Why should only those creatures be granted moral consideration whom we humans assume are capable of adopting the perspective of all others? What does the capacity for doing this have to do with moral considerability? What is the argument to show that the former be a condition for the latter? Why is this capacity the morally salient one among all sorts of capacities? And to move to the next stage in Habermas' argument, why does there have to be a "yes" or "no" response forthcoming from all those affected by a norm in order to assess its validity? That for lack of linguistic capacities such a "yes" or "no" will not be articulated, say, by the animals affected by some action, is trivial. But why grant such moral importance to the fact of the lack of this capacity? And does it appear plausible to proceed from this, as does Habermas, to the suggestion that "it cannot be determined" whether such a norm as "refrain from inflicting pain" would be accepted by the animals affected by it? Is it so that for lack of linguistic feedback from animals subject to the infliction of pain, we humans are simply agnostic about whether they want protection from it or not? Would it be paternalism on the part of us to decide, on behalf of the animals, that such pain is wrong and should be avoided? If so, is such paternalism more problematic than the attempt it entails to prevent what is seen as wrongdoing against animals who cannot "speak for themselves" the way we humans – backed by a moral theory some of us are committed to – demand?

Habermas goes on to point out how constitutive for sociocultural forms of life are "social interaction mediated by the use of language oriented to mutual understanding"; the person can develop an inner life and achieve a stable identity only to the extent that "he also externalizes himself in communicatively generated interpersonal relations and implicates himself in an ever denser and more differentiated network of reciprocal vulnerabilities" (1993: 109). The upshot is that "nobody can preserve his integrity by himself alone"; such individual integrity "requires the stabilization of a network of symmetrical relations of recognition", showing up morality as "aimed at the chronic susceptibility of personal integrity implicit in the structure of linguistically mediated interactions" (ibid.).

Let us grant here that this is a valid description of how human persons, in their individual vulnerability, depend on membership in a community with others where they are recognized; only in a social setting allowing for symmetrical relationships can each and every person become a truly individual such and be protected in his or her integrity. That these are genuinely social tasks and accomplishments that no individual can attain or preserve by him- or herself alone is an invaluable insight that goes back to Aristotle. The question is what happens when this model, conceived for humans-among-humans, serves as the standard for non-humans such as animals.

Habermas argues that in the case of humans, the personal integrity "implicit in the structure of linguistically mediated interactions…is more deep-seated than the tangible vulnerability of bodily integrity, though connected with it" (1993: 109. Why is this important? Because this difference between personal and bodily integrity "disappears in the animal world"; it does so "because we do not attribute personality to creatures with whom we cannot speak and cannot come to an understanding about something in the world" (ibid.). But again, why should we need to be able to speak with an animal in Habermas' linguistic sense in order to attribute personality to it? And why does recognition of the difference between personal and bodily integrity depend on our sharing the gift of spoken language with the animal in question? Can't we get at what is individual, about a particular animal without the mediation, or access, of spoken language? As Habermas himself maintains, these are capacities, and (social-linguistic) practices characteristic of and indispensable to humans. If the specific, capacities-based standard that applies to us does not fit the case of animals, so what? Doesn't what characterizes us, uniquely at that, according to the argument made, also show up an evolutionary selectivity about us, demonstrating not only what we are and can do, but also – by the same token – what we are not and cannot do? If we cannot access and understand animals the way we can other humans, isn't that a specificity and a limitation that help define *us*, not the animals?

Habermas acknowledges that we do in fact communicate with animals in a different way from that just described "once we involve them in our social interactions, in however asymmetrical a fashion" (1993: 109). He mentions our association with domestic animals as a case where our conscience is particularly insistent; it is so "concerning species with which we can communicate with particular ease", he quotes Panzig maintaining (ibid.). But again, isn't this to allow the way animals are treated to be determined by features to do with us, not features to do with the animals? Granting that some non-human creatures participate in our social interactions, we "encounter them in the role of an alter ego as an other in need of protection;

this grounds the expectation that we will assume a fiduciary responsibility for their claims" (ibid.). This is as far as Habermas is willing to go: "There exists a quasi-moral responsibility toward animals that we encounter in the role (if not *completely* filled) of a second person, one whom we look upon as if it were an alter ego. We can adopt a performative [as distinct from objectifying] attitude toward many animals, though not toward plants" (1993: 110). Within the interactions between us and the domestic animals Habermas has in mind, we must be able to "ascribe characteristics of agents" to them, among others "the ability to initiate utterances and to address them to us". In this situation, "we have duties that are *analogous* to our moral duties, because like the latter they are rooted in the presuppositions of communicate action. Of course, they are analogous to moral duties only to the extent that the asymmetries in the interactions admit comparison with relations of recognition between persons" (ibid.). The bottom line is that, to the extent that animals can enjoy moral protection from us, they may do so "only within the intersubjective horizon of *our* modes of interaction. Human beings always find themselves already within this horizon and as persons can never leave it, whereas animals belong to other species and other forms of life and are integrated into our forms of life only through participation in our interactions" (1993: 111).

In discussing the moral status of animals, we must be careful not to conflate what I maintain are two separate issues (as argued earlier): one has to do with what it takes to *enact* responsibility, and in this sense to qualify as a full-fledged moral agent. Here Habermas, as we have seen, holds the partaking in linguistically mediated interactions to be the decisive criterion: moral agents are agents possessing communicative competence, meaning agents capable of raising and responding adequately (i.e., rationally, by way of citing reasons) to validity-claims. The other issue has to do with what it takes to be a moral addressee – that is to say, at the receiving end of responsibility as exercised by moral agents. To *not* acknowledge that there is this distinction would mean (in the version I shall consider here, relevant to Habermas) to collapse the issue of what agency requires into the issue of what being an addressee requires, thus *making the former the condition for the latter.*

On the basis of the account of Habermas' discussion of animals above, where does he stand on this issue?

There are several things to take into consideration in arriving at an answer.

First, notice where Habermas from the very start places the burden of argument: namely with anyone who, like Patzig, advocates that the moral domain be *extended* from including only humans to comprising animals as well. In placing the burden one-sidedly on those who dissent from an anthropocentric framework, this framework is taken for granted as the point of departure, and the point of reference, for the discussion; hence there is no burden of argument in the other direction, for *limiting* the moral domain to humans only in the first place.

Second, to take the anthropocentric framework as a given in the sense of not questioning it, only the challenge to it, means that the discussion centres on the extent to which "new" candidates (such as animals) for inclusion in the moral domain possess the same, or relevantly similar, capacities to those already included, so that the more they do so, the better their chances of being admitted. In this way, as Levinas would say, otherness is reduced to the same; genuine otherness – from humans taken as the standard – counts against inclusion and operates as a reason why they must be excluded.

Third, the formalism of the approach taken by Habermas, unmistakably Kantian at that, presents it as a context-free, universal stance, enabling us to settle the issue of moral status *sub specie aeternitatis*, by reference to capacities-related criteria such as linguistic competence. This masks the *historicity* of this stance, the fact that "the denial of thought, feeling and consciousness to nonhuman animals" that Habermas subscribes to (albeit not in the crude version of Cartesian animals as "mere automata") is a historically recent phenomenon, culturally specific to the West. B.E. Rollin, in his aptly titled article "How the animals lost their minds. Animal mentation and scientific ideology" (1990), shows that before the turn of the twentieth century there was a strong belief in in both animal cognition and emotion by scientists and lay people alike. As we saw in Chapter 1, Darwin highlighted an evolutionary continuity with animals and humans for both emotions and cognition; differences are differences in degree, not in kind. Rollin holds that the ontological shift away from this knowledge was due to a reductionist trend in society in general, with a new "emphasis on sweeping away frills, excesses, superfluities" (quoted in Kessler 2019: 242). As Caroline Merchant demonstrates in *The Death of Nature*, the Enlightenment joined forces with the scientific as well as industrial revolution from the seventeenth century onward to embark on a course that would advance the desire to take from more-than-human nature whatever was wanted, doing so with impunity and for the sake of "progress" and without obligation to the countless non-human species exploited for human needs and ends (see Kessler 2019: 243).

More specifically, Habermas' neglect of historical change and of how the contexts for humans' experience of animals differ, both between cultures and from one epoch to the next, is betrayed in his statement that "we can adopt a performative attitude toward many [i.e., domestic] animals, though not toward plants" (1993: 110). Who is this we? Hardly a universal and timeless one. While we would have to go way back in pre-modern time in the Western society Habermas probably alludes to find a performative attitude to plans, such an attitude is an important part of everyday life among many indigenous peoples in the world today, though under enormous pressure to abandon such practices (dismissed as based on superstition) and replace them with the modern scientific world view theorized by Max Weber and of which Habermas is a contemporary champion. (I have discussed the persistence of the Cartesian view of nature in Habermas' overall theory of modernity in Vetlesen 2019.)

Fourth, while many would agree that linguistic communication is a crucial part of the interaction instrumental to the development as well as protection of human identity and integrity, Habermas simply presumes, rather than argues, that such communication is to be regarded as relevant to relationships between humans and animals – only to turn his not finding it there against the animals, as proof of something humans have that they lack. Why should that be important, for whom, and with what sort of implications for the issue under discussion, the moral status of animals? Reading Habermas' reply to Patzig, it is striking that virtually everything that is said about animals concerns what they (presumably) can't do, when compared to what humans can do, that is. Little is said about what animals *can* do, and nothing at all about what (some) animals can do that humans can't. As part of this, there is a certain nonchalance in making very general claims about animals without distinguishing between wild, domestic, and pets; and in particular, in not bothering to consult the research carried out in animals' studies before stating, matter-of-factly, that "an animal does not experience its pain reflexively like a human being, who in suffering is cognizant of the fact that he is in pain" (1993: 110). Does this really hold for, say, a wolf, a cat, and a pig in pain, to pick an example from each of the three categories mentioned? Besides being problematic as a descriptive characterization of how "an animal" falls short of experiencing pain in the complex ways humans do, the statement seems to encourage the normative conclusion that, for this reason, the animal's pain is less worthy of moral consideration.

Fifth, consider the way that Habermas makes claims for moral consideration dependent on being capable of "adopting the perspective of all others". Even if we grant that this is a crucial ability in morally relevant interactions between humans, there is little in terms of attempting to adopt the perspective of animals (or some particular such) in Habermas' text. And whereas "adopting the perspective of all others" is certainly a tall order in most human communities, the prerogative of the tiny minority that ascends to the post-conventional stage of the Kohlbergian model Habermas' relies on in his discourse ethics (see 1990), and as such presumably not present at all in animals, including mammals like chimpanzees. Studies carried out in, say, South America are fraught with accounts of how adept various species of animals are in adopting the perspective of their counterparts, not only other animals at that – their ability to put themselves in the shoe of the humans hunting them and anticipate their next move is the stuff of which their daily life is made. In existential terms, and given the increasing physical proximity imposed by humans on the animals (as exemplified in the corona pandemic), they are no less dependent on the ability to read and adjust to their species others than are humans – perhaps more so.

Sixth, there is Habermas' choice to focus on animals that participate in our interactions, observing that they can "enjoy moral protection only within the intersubjective horizon of *our* modes of interaction" (1993: 110). Given this focus, the animals Habermas has in mind are domestic ones, such as cats or dogs; this is as concrete as he gets. To be sure, it seems reasonable to hold that animals that we have chosen to possess and that form part of our "expanded" family are entitled to our protection, more so than are animals we have never encountered, let alone developed a close and long-lasting relationship with. In other words, we need to differentiate between animals that are "integrated into our forms of life…through participation in our interactions" (1993: 111), and all the animals that do not and never will. Recall that in saying this, Habermas comments: "The limits of our quasi-moral responsibility toward animals are reached once humans, in their role as members of one species, confronts animals as exemplars of another" (ibid.). Why does he say this? The way I read it, his claim is that in the moment we humans confront animals as our "species other", our quasi-moral responsibility towards them ceases.

This is highly problematic. As it stands, the claim entails that moral responsibility as incumbent on humans is confined in its reach to species kin: other human beings. Membership of another species is tantamount to being, and having to remain, on the outside of the moral domain.

But I doubt that Habermas means it as categorically as that. Preceding his comment, as we saw, is the attention he pays to animals as "integrated into our forms of life…through participation in our interactions". Species membership does not *per se* determine the matter, in the form of limiting the reach of responsibility to fellow humans only. There is a gradualism involved in that, insofar as *some* animals participate and others do not, those that do are entitled to a moral consideration that the others are not.

Of course, I have already said as much. But only now can we appreciate the full context. What I find highly problematic is the way Habermas here appears to give moral weight to a factor that is wholly contingent – namely, whether or not an animal participates in interactions with us. Why should *that* be a morally salient factor? One could respond that animals we form relationships with are animals we come to love, precisely in their individual particularity (Habermas hesitates to use the word "personality"); and from this emotionally charged closeness springs a type of moral responsibility on our part that is "naturally" reserved for this particular animal, as such denied all others.

But such a closeness-oriented ethics, giving pride of place to the emotional impact of forming a relationship with a particular "other", does not at all square with Habermas' own moral theory, being post-Kantian and deontological; in fact, it sits much better with one of its main rivals, the Carol Gilligan-inspired "ethics of care" (1982). *Impartiality* as conceived and endorsed by Habermas demands that the contingency of knowing or not knowing, loving or not loving the other, be abstracted from when adopting the proper moral point of view. Habermas offers no argument to justify why the contingency he denies moral import among humans should make a moral difference with regard to animals.

However, and seventh, there is another aspect to this, more subtle and important. Let me put it like this: Habermas' discussion with Patzig is an exemplary exercise in *methodological and moral individualism*. In taking as his framework from start to finish the relationships with humans that an animal may be part of, Habermas restricts his analysis to that of one-to-one encounters. Though admittedly pedagogically attractive, it is striking that Habermas (like Kant before him, with his predilection for horse and dog examples) chooses to concentrate on the issue of the individual moral agent's responsibility for individual animals. Why is this important? In advancing his argument within that setting, Habermas misses out on the pivotal issue at the core of holistic theories concerned with the conditions under which animals currently live – namely, how to make sure that they thrive and persist as the other-than-human life forms they are, rather than being exploited to the point of decimation and extinction. An ecologically informed holistic approach to the animal question has to acknowledge that when the whole (the ecosystem and biosphere of which animals are a part, taking the form of habitats and biotopes in the concrete, local case) is under threat, each individual is. In nature the individual animal can only survive provided its ongoing interaction with other individuals, with members of other species (predator-prey), and with its total environment is thriving; and these inter- and cross-species interactions can only be intact provided the whole (the ecosystem, say, the forest) they help make up, and within which each species occupies an indispensable role, is and remains intact. To be sure, we may want to argue, in keeping with methodological and moral individualism, that the individual tiger has intrinsic value for what it is in itself (per analogy, if you will, with Kant's position on the inviolability of the human person). But once removed from the jungle, this intrinsic value, conceived as independent of the tiger's wider environment, its reliance on so many "others", would collapse. Therefore, the moral questions – of agency, of status, of responsibility, obligations, and rights – must address the habitat that every single exemplar is dependent upon: if you want to save the tiger, you need to save the jungle. There is no way that methodological individualism can be permitted to translate into moral individualism here, so that the individual would trump community, the exemplar the (eco) system, in the sense of the former being self-sufficient in its possession of intrinsic value vis-à-vis the latter. Indeed, the juxtaposition in such a version is entirely misguided within an ecological framework, meaning *factually* false: no thriving and value-able whole, no thriving and value-able individual. The conditions of existence ecologically understood are paramount; any moral theory that concerns itself with creatures – be they humans, be they animals – whose lives depend on nature needs to take these conditions as its point of departure.

To be sure, this is a way of framing the issue of the moral status of animals that must appear totally foreign to the approach pursued by Habermas – yet it is a commonplace in environmental ethics as elaborated by prominent figures such as Paul Taylor, J. Baird Callicott, and Holmes Rolston during the last four decades. My point here is not simply that Habermas discusses the animal question in a way that omits ecology; perhaps he thinks that doing so is part of an intellectual division of labour, allowing for an assessment of the moral status of animals that can be carried out abstractly, concentrating on procedures and on the validity of norms, as Kant-inspired ethics is wont to do.

But such a separation between environmental and animal ethics will not do. The moral questions to do with animals must address the habitat that the physical existence of every single individual, regardless of particular species, depends upon; neglecting this cross-species dependence is a non-starter in theory no less than in real life. Try, and the consequences are but a lesson in how to self-destruct. To acknowledge this is to realize that there is something reductionist, verging on atomistic, in the way Habermas removes (abstracts) animals from their generic form of existence in a complex interplay with a large number of other species in nature; removes them, that is, from the whole of which they are parts (participants) and outside of which they cannot live, only to let them – one by one – become a topic for ethics and a concern of practical moral import the moment they come into contact with specifically *human* forms of life, as not-so-fully-capable partners in *our* networks of interaction.

Yet it is not only from an ecological point of view that Habermas' failure to situate animals within their proper life-context in their natural habitats is problematic. Removing them by way of abstraction from their natural environment, whether it be the forest, the lake or the sea, amounts to imposing a veritable limitation: in making animals' entry into "our networks of communication" the criterion for discussing them, Habermas leaves out of consideration all animals that are not met upon by humans within those networks. In other words, animals are effectively restricted to domestic such, exemplarily to pets like cats and dogs. For lack of partaking of relationships with humans, non-domestic animals of all sorts are delegated to a kind of analytical no mans' land, with the consequence that the question of *these animals'* moral status, as bearing on their right to protection, is simply not addressed. This is a high price to pay for restricting the framework for discussion, and so the entities to be considered, the way Habermas does. Clearly the types of animals with which we do *not* interact the way we do with dogs or horses put the issues of environmental and animal ethics alike into much sharper focus than do domestic animals. It is precisely the element of *otherness* exhibited in, say, a tiger or an eagle owl that commands our attention, an otherness tied to these animals' *absence* from our networks of interaction, and one that crucially depends on the tiger's or eagle owl's non-assimilation to our current life form, increasingly urbanized and remote – mentally as well as geographically – from what remains of their original, natural habitat. This being so, a major issue in an ethics concerned with animals today is to do with the importance of ensuring that wild animals, animals in the wild, will have a viable future, since it can no longer be taken for granted that they will. It will require concerted efforts to secure the future existence of wild animals, relying as it does on successful efforts to protect the natural habitats from which their lives are inseparate, habitats that today are being fragmented, spoiled, and destroyed by human activities, causing the animals to be decimated or even to go extinct not as a consequence of one-on-one encounters with humans (Habermas' favoured framework) but as unheard, unseen casualties of the *modus operandi* of silent killing.

The historicity and selectivity witnessed in Habermas' approach to animals justify seeing it as a case of shifting baseline syndrome: Habermas takes for granted the condition of animals, and the nature of their relations with humans, as they both appear at a particular moment in history. It thus becomes "natural" to concentrate on some animals rather than others. One of the many ways in which anthropocentrism is taken for granted as both point of departure and framework is that animals of the kinds that humans nowadays meet upon within their lifeworld, are the animals considered as candidates for moral consideration. That some animals are within this human-centred lifeword, others outside it, is the result of a long historical development, in the course of which animals – generally speaking – have become decreasingly wild, increasingly domestic, rendering the sheer future existence of those still wild, or living in what is still wilderness, precarious in a way that is truly unprecedented. The impact on humans-animals relations as well as on the plight of animals involved in the historical changes we are talking about is wholly neglected in Habermas' discussion with Patzig.

Before concluding, however, there is another discussion to take into account, where Habermas addresses at least one of the issues I just faulted him for neglecting. In his reply to another colleague, Henning Ottmann, he considers extending ethics "beyond the domain of interpersonal relations to our relationship with creatures that cannot fulfil the conditions for responsible action" (1982: 247). This sounds promising. Significantly, however, Habermas goes on to frame the issue within what he calls "the ecological problematic", observing that "with the heightening of our technical power of disposing over nature, there also comes increased responsibility for the natural processes set in motion by our interventions". He then goes on to say: "This problematic can be dealt with satisfactorily within the anthropocentric framework of a discourse ethic" (ibid.). Well, can it? What justifies Habermas' confidence?

Habermas considers the ethics of animal protection that advocates the extension of "one's neighbors" to "all affected living creatures with whose suffering we can empathise", the intuition being that "all those creatures rescued by Noah's Ark *should* enjoy the protection of subjects with whom we interact on the level of equals" (1982: 248). He comments that attempts to open up "a *moral access* to nature-in-itself are by no means absurd"; nonetheless, he warns that we should not "permit ourselves to be cajoled by such intuitions into ignoring the difficulties we encounter here, just as with the analogous attempts to gain *theoretical access* to nature-in-itself" (ibid.).

It is not clear why Habermas moves the way he does from "affected living creatures", animals included, to the question of gaining access to "nature-in-itself". If not a case of changing the subject, the move has the effect of casting the question in the manner of Kantian metaphysics and Kantian epistemology: the in-itself attributed to nature here abstracts from the living creatures whose suffering is plain to see (experience, relate to) in keeping with the intuition mentioned above; nature-in-itself mystifies the suffering *of* the animals as experienced – without doubt or epistemic qualifications – *by* us. *This* is the very phenomenon that gives rise to the intuition behind the sort of ethics Habermas purports to consider. Shifting the issue from its original context onto a Kantian inventory where matters of method (access) have primacy over subject matter is of little gain as far as doing justice to it qua moral issue is concerned.

But Habermas also offers a more specific argument: "An ethic that gets rid of its 'anthropocentrism' extends the circle of addressees lower in the scale of nature; but again, the limits of empathy into alien life remain blurry. And is it not the case that the precepts of sympathetic solidarity soon come to a halt – namely already with plant life – if they are not to come into conflict with the firmer imperatives of the self-preservation of the human race?" (1982: 248).

As Habermas must know, this is a question, not an argument. Yet we need be in no doubt about where Habermas stands. The case for solidarity with non-human species and life forms comes up against a limit once it comes into conflict with "the firmer imperatives of the self-preservation of the human race" (ibid.). To be sure, Habermas is perfectly right in making explicit that this is what anthropocentrism in ethics in fact implies. But what we want to know is not simply what this position *entails* with regard to the status of animals, or plants. We want to know whether the position is *convincing*.

Rather than providing a separate argument, Habermas states that "the basic philosophical question" here is to do with how an ethic of animal protection (or generally, a non-anthropocentric ethic) "could be grounded at the level of learning attained in the modern understanding of the world" (ibid.). What is he referring to? Well, it is the kind of world view that Habermas, following Max Weber, terms modern and scientific, replacing the "mythical" world views of which, say, an animist cosmology like the one discussed by Rose in the previous chapter would be an example. The "disenchantment" of the world Weber spoke of entails that nature *in toto* is denuded, is rid of so-called occult forces of all kinds, that is to say, everything psychic and mental, everything exhibiting interiority and subjectivity, feelings and intelligence, purpose and value; in the form of a sharp ontological divide, if not an abyss, the latter are regarded as qualities the sole possessors of which are human beings.

There is no need to go further into Habermas' debt to Weber's distinction between mythical and modern world views (see Habermas 1984, 1987; Vetlesen 2015). What matters here is that in subscribing to Weber's notion of disenchantment of "external" (i.e., non-human) nature, Habermas maintains that for this domain of reality there is "only one theoretically fruitful attitude, namely the objectivating attitude of the natural-scientific, experimental observer" (1982: 243f.), facilitating the interchangeability of the (human) observer (scientist), the repeatability of the experiments/observations carried out in a controlled setting (exemplarily: the lab), and the replicability of the results obtained. In other words, everything to do with the particularity of the human participants, and with the particularity of what can be encountered "in nature" given the specifics of the spatio-temporal setting – that is to say, everything that goes to make up the local knowledge characteristic of indigenous cultures' "mythical" world views – is abstracted from in the model that Habermas takes over from Weber and from Bacon and Descartes before him. The conditions of instrumental action of a sort requiring an objectifying attitude, enabled by adopting a detached, neutral/impartial stance to it, Habermas maintains, "bind our knowledge of nature to the interest of possible technical control over natural processes"; labour, understood as Marx did in *Capital* as the behavioural system of instrumental action, is the "invariant relation of the [human] species to its natural environment" (1972: 36, 309). This being so, attempts to "use the experimental potential gathered in non-objectifying dealings with external nature for purposes of knowledge" would, Habermas contends, "lead back to metaphysics, and thus behind the levels of learning reached in the modern age into a re-enchanted world" (1982: 245).

Thus Habermas repeats his warning against falling behind "the level of learning" attained in a Western modernity based on disenchantment. There seems to be, for him, something sacrosanct about this learning. While it would be foolish to deny that the science-based world view he espouses has brought genuine progress in many areas of life, it would be disingenuous to deny its complicity in the exploitation of the natural world that is the "dark side" of the progress reached, a side that Habermas is conspicuously less interested in investigating than were his predecessors in Critical Theory, Adorno, Horkheimer, Benjamin, Fromm, and Marcuse.

To end this section, consider Habermas' aforementioned statement that "the precepts of sympathetic solidarity soon come to a halt – namely already with plant life – if they are not to come into conflict with the firmer imperatives of the self-preservation of the human race" (1982: 248). Why is there this conflict? Why should a stance of solidarity with non-human species – be it plants, be it animals – need to conflict with the preservation of humans? What kind of society has the better historical record when it comes to preserving, as opposed to overexploiting and destroying, the natural world on which the preservation of human life depends? Why indeed is conceiving of, and acting upon, the nature-humans relationship as one of conflict superior to its historically enacted alternative, that of the interplay between and across species within what animists term – and experience as – a comprehensive ecology of selves, so many different selves? Does Habermas' notion of conflict acknowledge the fundamental ecological insight that species are inter-dependent and that each one of them – humankind not excluded – can only thrive if others do?

Habermas does not seem willing to recognize this insight and its implications. Indeed, the fact that humans remain just as dependent on nature in the present era where the "scientific" world view reigns as in the pre-modern era, is not heeded in his theory. Ultimately, the possibility of a non-instrumental and non-exploitative relationship with non-human creatures such as animals championed by Benjamin and Adorno is ruled out for conceptual reasons in Habermas' oeuvre from start to finish. Why? Because on his view the modern type of mastery over nature is "a product of the human species as a whole, and not of an individual epoch', a specific class or a surpassable situation"; thus the "transcendental necessity" of instrumental action aimed at technical control over nature, making for what he considers the "invariant" relationship between humans and nature (1971: 88). This being so, Habermas' concern with falling behind insights already attained can be turned against him: that the human species in its self-ascribed role and entitlement as nature's master is itself dependent on nature is a seminal insight in his predecessors' work *Dialectic of Enlightenment*: the science and technology driven attempt at all-out mastery over nature may well turn against the aggressor. In the words of Henning Ottmann: "If it is true that progressive mastery over nature possibly results in the destruction of nature as the foundation for survival, one can very well claim that nature itself is the subject's master whose 'imperatives' the subject had better 'obey'" (Ottmann 1982: 92). A nature that *transcends* the attempt to master it, such as to *resist* the "interest in technical control" postulated by Habermas on the level of the human species, is a nature of a sort simply not conceptualized in his theory: external nature as he conceives it simply *is* the realm of lifeless matter that can be manipulated without limit. Thus there is a perfect *fit* between technical control, instrumental action, and objectifying stance on the one hand, and nature as the object domain that *in toto* corresponds to them, on the other. As becomes increasingly evident, to continue to act on the basis on this assumption is a recipe for ecological disaster.

II

That humans' capacity for autonomy as understood by Kant has both set us apart from the rest of nature and helped justify our dominance over it is something that has never attracted Habermas' critical attention like it did Adorno's.

In *The Animal That Therefore I Am*, Jacques Derrida seizes upon this aspect of Adorno's thinking, reminding us that according to Adorno the capacity is "directed against animals" (2008: 100). The autonomy-based sense of entitlement in seeking mastery over nature by means of science and technique contains a considerable element of aggression, albeit one that is typically not admitted as such but hidden behind the rhetoric of historical progress. The brutal truth that Derrida credits Adorno with insisting upon is that the mastery sought is nothing short of "an act of war and a gesture of hate, an animosity, as though Kant were raising the stakes in a bid to add venom to a Cartesian project that, for its part, would remain essentially neutral and indifferent, fundamentally different to the animal machine" (2008: 100f.). Derrida goes on to add that he for one thinks that "Cartesianism belongs, beneath its mechanist indifference, to the Judeo-Christiano-Islamic tradition of a war against the animal, of a sacrificial war that is as old as Genesis". That violence or war, he continues, "has until now been constitutive...of the very possibility of techno-scientific knowledge within the process of humanization or of the appropriation of man by man, including its most highly developed ethical or religious forms. No ethical or sentimental nobility must be allowed to conceal from us that violence, and acknowledged forms of ecologism or vegetarianism are insufficient to bring it to an end, however more worthy they be than what they oppose" (ibid.).

Derrida does not tell us what would be required to bring this violence to an end. I think that his not doing so is the price he pays for locating it everywhere he looks for it, be it in Cartesianism, Kantianism, or the comprehensive religious tradition he alludes to. By locating the violence so deeply and widely, Derrida risks casting it as virtually inseparable from humans' relationship with animals throughout modern history. It must be said that Adorno is guilty of the same tendency, inspiring commentators such as Deborah Cook to speak of nature as what humans have always been busy "othering", so that "whatever nature is said to be, we have stubbornly defined ourselves as both superior to nature and its opposite" (2011: 46). Formulations like this fly in the face of the fact that for the animistic cosmologies that have been dominant for the greater part of humans' existence on earth, relations with animals and other non-human species have been perceived as pervaded by continuity and kinship. This is not to deny the existence of violence against animals in premodern times and cultures, a violence that would be denounced should it be considered excessive or gratuitous and so failing to display the respect seen as due to animals in the cosmologies in question.

On the other hand, Derrida is on the right track when he observes that the violence against animals Adorno sees epitomized in Kant is one that takes the form of hate and revilement against animality. What sort of animality, in whom? The obvious answer is the animality that marks animals as non-human, as our "other", the animality cited by the "bellicose hatred in the name of human rights", that "far from rescuing man from the animality he claims to rise above, confirms the waging of a kind of species war and confirms that the man of practical reason remains bestial in his defensive and repressive aggressivity, in his exploiting the animal to death" (2008: 101). Derrida goes further still, suggesting that "bad will, even a perverse malice, inhabits and animates so-called good moral will; and that this 'evil', the malady and malignity of this evil, is borne not *in* and *on* the animal, but *against* the animal, to which it is a question of doing and wishing harm" (ibid.).

Whence this "Kantian or idealist hatred of the animal, this zoophobia"? Adorno tries to explain it by pointing to the "resemblance or affinity between human and animality" ("die Tierähnlichkeit des Menschen"); nothing is "more abhorrent, more hateful, more odious to Kantian man" than the memory of this resemblance (Derrida 2008: 103). For this reason, Kant allows no place for compassion between human and animal in his ethics. Thus the animality of the human becomes the great taboo for Kantian man (yes, man, not woman). That this should be so is in keeping with Adorno's overall assessment of idealism and transcendentalism á la Kant, holding the project of human mastery over nature and over animality to be absolutely crucial to it. Not content to leave it at that, Adorno takes his criticism much further – he alleges that (again in Derrida's paraphrasing) "for an idealist system, animals virtually play the same role as Jews did for the fascist system". Such a fascism "begins whenever one insults an animal, even the animal in man"; accordingly, "authentic idealism consists in *insulting* the animal in the human or in treating the human as animal", whereby the insult doesn't just imply verbal aggression, but "an aggression that consists in degrading, reviling, devaluing someone, contesting his or her dignity", recognizing that insult is directed at *someone*, not something (ibid.). Having already extended the idealist hatred of the animal to hatred of the Jew, Derrida remarks that one could easily extend it further, to a hatred of femininity, or of childhood. He concludes that insulting the animal "would therefore be a fact of the male", so that "the animal's problem [*mal*] is the male", borne out historically in that "from hunt to bullfight, from mythologies to abattoirs, except for rare exceptions it is the male that goes after the animal, just as it was Adam whom God charged with establishing his dominion over the beasts" (2008: 104).

Regrettably, Derrida is content to paraphrase Adorno's criticisms of Kant; he does not engage with Adorno's own position. He thus misses the opportunity to dwell on the claim that what triggers the hatred of animals is *their affinity to the animality in humans*, rendering the aggression against animals a sort of self-hatred. Without going deeply into Adorno's thinking here, for him the domination over what Habermas calls "external nature" is inseparable from the suppression of man's "internal nature". The elevation of rationality as proof of human exceptionalism and as the paramount criterion that secures humans, and humans only, moral status, entails that the "non-rational" features count as something negative, devoid of worth, a view seen as corroborated by their exhibition in animals. Hence the taboo against the animality of the human that we saw Adorno point to in Kant works in two directions, as it were: on the one hand, in portraying the animal-like in humans as what detracts from and is at odds with what endows us with moral status, it denies that *our animal nature essentially belongs to our humanity and to that in virtue of which we are entitled to moral consideration*; on the other hand, in portraying what we share with animals, and what animals share with us, in this negative light, perceived similarities between us and animals become themselves something negative, a reason for keeping our distance from animals. What we cannot tolerate in ourselves, we will not tolerate in our species others. Jay Bernstein puts this well when he states that "because modern moral thought for the most part eschews animal nature as the fundament of meaningfulness, it must eschew the roles of pain, suffering, damage, and injury as pivots for moral thought and replace them by some broader notion of moral wrongness"; hence "it is not surprising that suffering should play such a marginal role" (2001: 198).

Adorno had a sharp eye for the many ways in which a trace of animality is regarded as a "matter out of place" in contemporary society, citing the outcry that the smell of sweat or the sighting of hair under a woman's armpits now cause. What is animal-like is what is dirty, raw, primitive, exposing what should be hidden or simply removed as unbecoming of fully socialized individuals who have successfully and responsibly accomplished the transition from natural to cultural beings, a transition in the name of human dignity and freedom yet bought at the price of dictatorial repression of the entire intrasubjective dimension seen as their threat. As Michael Pollin observes, this discomfort with our animality is being transferred onto the animality of animals as well, as exemplified in "the anguished handwringing in some animal rights literature over the fact of predation in nature" (2006: 321), a tendency I identified in Cynthia Willett's work in Chapter 2.

Rather than dwelling on the taboo against animality, Derrida pursues his reflections on animals by engaging with Levinas. Even if Levinas submits the subject of ethics to a radical heteronomy that stresses receptivity and passivity in the sense that the subject is not "the first mover" but instead is being moved by way of encountering the Other, the face that issues the command "thou shalt not kill!" remains "first of all a fraternal and human face", that of my brother or my neighbour. Derrida states bluntly that "it is a matter of putting the animal outside of the ethical circuit", so that, if the human subject is for Levinas a face, "according the animal or the *animot* any of the traits, rights, duties, affections, or possibilities recognized in the face of the other is out of the question" (2008: 106). That this should be so may come as a surprise because, after all, "isn't the animal more other still, more radically other, than the other in whom I recognize my brother"? But no. "It seems precisely that for Levinas the *animot* is not an other". Indeed, the issue seems fully settled, leaving no doubt as to Levinas' position: "The animal", writes Derrida, "has neither face nor even skin in the sense Levinas has taught us to give to those words. There is, to my knowledge, no attention ever seriously given to the animal gaze, no more than to the difference among animals, as though I could no more be looked at by a cat, dog, monkey, or horse, than by a snake or some blind protozoon" (ibid.).

Does the animal have a face in the sense that one can read "Thou shalt not kill" in the eyes of the animal? We discussed this question when examining Deborah Bird Rose's reading of Levinas' comments on Bobby above. Derrida quotes the response Levinas gave to John Llewelyn in 1986: "I cannot say at what moment you have the right to be called 'face'. The human face is completely different and only afterwards do we discover the face of an animal. I don't know if a snake has a face. I can't answer that question. A more specific analysis is needed" (Levinas quoted in Bernasconi and Wood 1988: 171f.). Having said this, sounding agnostic, Levinas adds:

> It is clear that, without considering animals as human beings, the ethical extends to all living beings. We do not want to make an animal suffer needlessly and so on. But the prototype of this is human ethics. Vegetarianism, for example, arises from the transference to animals of the idea of suffering. The animal suffers. It is because we, as human, know what suffering is that we can have this obligation.
>
> (Levinas quoted in Bernasconi and Wood 1988: 172)

Derrida makes much of Levinas' remark about the "prototypical" character of ethics as human, taken as "the space of a relation between humans, only humans", implying that it is only by means of a transference that we become sensitive to animal suffering (2008: 108). In particular, he takes issue with Levinas' statement that "It is because we, as human, know what suffering is that we have this obligation", i.e., that we do not want to make an animal suffer needlessly. But Derrida does not tell us what exactly is wrong about Levinas' recourse to our own (human) exposedness to suffering as the experiential basis for our obligation to protect animals from suffering, nor does he offer an alternative. Would Derrida object to a statement to the effect that exposedness to suffering is a basis for the obligation *among humans* to protect against suffering? Why is it mistaken to identify the source of moral consideration in others, of their being *in need of it*, by virtue of identifying in oneself that same source, eliciting the same need? Or is the problem Derrida sees here a different one, having to do with giving such a prominent role to vulnerability and suffering? Should the basis for ethical obligation – be it for the human other, be it for the animal – be located elsewhere? In a different capacity? Or in some other source or parameter altogether? Is the mistake that of attempting to find (or establish) *the same source* for obligation to both humans and animals? Should there rather be two different sources, one for each ethical "other"? Does the source have to be other-than-human when the addressee of the obligation is other-than-human? In that case, is Levinas' mistake that of making a move from the same to the other? Does doing so belie his ambition to offer a genuinely other-directed ethics, one where the other's difference from the self is what counts, what the obligation is all about?

It is unfortunate that Derrida omits examination of an ethics of compassion, a kind of ethics that he observes Adorno misses in Kant, leaving it at that. This is not to suggest that Levinas should be read as espousing such an ethics (he does not), but exploring it would have had the advantage of allowing Derrida to clarify his views about the moral significance of suffering and vulnerability, considered as experiences shared by humans and animals. Instead he focuses on Levinas' admission that he "can't answer that question", contending that "declaring that he doesn't know where the right to be called 'face' begins means confessing that one doesn't know at bottom what a face is, what the word means, what governs its usage, and that means confessing that one didn't say what responding means", which amounts to "calling into question the whole legitimacy of the discourse and ethics of the 'face' of the other" (2008: 109). Notwithstanding the indecisiveness voiced by Levinas' "I can't answer that question", Derrida for his part sees no grounds for being in doubt: Levinas' "nonresponse" is a "no", a "no" which "echoes everywhere, can be heard...as soon as Levinas chooses *an example* of an animal in order to support his declaration of nonknowledge" – the example being, as we saw, a snake, that is to say an animal with immense mythical weight, something that makes "attributing a face to this figure of temptation or evil highly improbable". For, proceeds Derrida, "where there is evil there is face. What remains faceless is pure indifference to good and evil" (2008: 110). I take Derrida to mean by this that to deny the other – any other – a face is to help justify the evil done to that other, to rob it of the authority and the resistance that only the possession of a face can muster. In choosing the snake Levinas avoids lighting on more disturbing examples, such as "the cat, the dog, the horse, the monkey, the orangutan, the chimpanzee – whom it would be difficult to refuse a face and a gaze", and hence "to refuse the 'Thou shalt not kill' that Levinas reserves for the face, for the face of the human for the human, or for God's commandment in instituting the nakedness of the human face" (ibid.).

I have quoted so extensively to helps us examine exactly what Derrida is doing. To begin with, the emphasis he puts on Levinas' knowing – or rather not knowing – precisely what qualifies as a face and what governs the use of the word flies in the face of Levinas' insistence that ethics is "first philosophy", that it is prior to epistemology and ontology, that is to say, pre-intentional and pre-cognitive, so that responding to the other in the form of responsibility not be premised on knowledge about the other – the other's identity, the other's plight. Ethics being prior to knowledge and the entire business concerned with what we can or cannot know means that responsibility is operative – is received, imposed – independently of its workings and its results in the particular case, that the latter *not* be regarded as a condition (presupposition) for it. Moreover, when Levinas says "I don't know if a snake has a face", he might have added two comments, both in keeping with his ethics:

one, that not "knowing" this about the snake it not what matters – the fact that the snake is vulnerable to suffering is, since sufficient to obligate me, to posit me as responsible; and two, that not being sure if a snake is a face does not entail being indecisive about whether a cat or a dog or an orangutan is a face – phenomenally, the snake is a less clear-cut case than the latter, and Levinas should not be faulted for admitting as much.

But Derrida isn't content to leave it at that. Asking what is forbidden by Levinas' "Thou shalt not kill", he notes that Levinas often translates it as "You shall commit no murder", i.e., it forbids homicide. This means that it doesn't forbid "putting to death in general"; it "doesn't forbid one to kill an animal; it forbids only the murder of the face", meaning, Derrida continues, "the face of the other, my neighbor, my brother, the human, or another human" (2008: 110). When putting an animal to death, or exploiting it to death, is not forbidden, this is so because the animal, "inasmuch as it is incapable of being the victim of a murder, doesn't die", only humans do, as Heidegger maintained (ibid.). Holding Levinas to leave us with an exclusive connection between the face, murder, and the human, Derrida once again wants to demonstrate that there is no place for animals. Yet he does so at the price of twisting what Levinas is actually saying: the species-neutral "Thou shalt not kill!" is Levinas' signature formulation of the command, and when he considers murder as a case in point he should be allowed to do so, without this implying that killing is wrong *only* when the victim is a human being. The point Levinas makes very clearly is that "the animal suffers"; knowing ourselves what suffering is, "we can have this obligation [to protect an animals from suffering needlessly]". Indeed, as Levinas says without doubt or qualification in the quotation that Derrida himself brings, "it is clear that, without considering animals as human beings, the ethical extends to all living beings" (2008: 108). Reference to this statement is conspicuously missing in Derrida's critical commentary.

However, Derrida does have something to say about Levinas' essay on Bobby. Like Deborah Bird Rose and so many others before and after him, he seizes on Levinas' calling Bobby "the last Kantian in Nazi Germany": "For him – there was no doubt about it – we were men" (2008: 114). Setting out to bring back down to earth "all the enchanted readers who dream of reconciling Levinas' ethics with the animals", Derrida deems it necessary to "limit straightaway the scope of this hymn to Bobby" (ibid.). His main criticism is voiced in his question: "But how can one ignore that a Kantian who doesn't have 'the brain needed' to universalize maxims would not be a Kantian?", to which he answers: "Bobby is thus anything but Kantian" (ibid.). Perhaps this is just as well, however. Why? Well, in light of "the cruel and merciless war that a virtual 'fascist' (in Adorno's terms) Kantian idealism declares on animal life, calling Bobby a Kantian is no compliment" (2008: 115).

On Derrida's interpretation, then, there seems to be no way to find reasonable what Levinas says about Bobby. Leaving aside for now Derrida's invoking of Adorno's charge of fascism (which is hardly representative of his assessment of Kant's moral philosophy; see Adorno 2000: 144), a more balanced reading of both Kant and Levinas would point out that when Levinas remarks that for Bobby "we were men", Bobby is ascribed the capacity to *recognize others*, in the sense of *Anerkennung* so crucial in German idealism; he is thus effectively recognized in Levinas' portrait of him as a moral subject, not in the full-blown sense, to be sure, but in the minimal, yet important one of making a difference to those he recognizes, the Jewish prisoners, as I pointed out above. Since according to Kantian premises you need to be a moral agent to qualify as a moral addressee, Bobby's act of recognition entails that he is fellow member of the men's moral universe, and precisely not excluded from it, as Derrida's reading of Levinas' remark has it. The other point worth making here, made in a slightly different manner in my earlier discussion of Rose, is that Derrida chooses to ignore the social context in which Bobby acts like he does. To assess whether calling Bobby a Kantian is a compliment or not is only half the story. To say of Bobby that he acts like the last Kantian is to say that the Germans at the scene do nothing of the sort: the other side of the compliment to Bobby is the condemnation of the German guards, brought into sharp relief by way of contrast with a dog. There is nothing here to the detriment of the dog, but everything to the detriment of the men. That they, presumably, qua humans, have the brains needed to universalize maxims only goes to underline their moral failure.

III

In his seminal essay "Why look at animals?", John Berger remarks that following the physical and cultural marginalization of animals in Western society since the seventeenth century, "animals are always the observed. The fact that they can observe us has lost all significance". He adds that "public zoos came into existence at the beginning of the period which was to see the disappearance of animals from daily life. The zoo to which people go to meet animals, to observe them, to see them, is, in fact, a monument to the impossibility of such encounters" (1980: 16, 21).

Without mentioning Berger's essay, Derrida writes that common to the grand forms of philosophical treatises regarding the animal is that they are authored by "people who have no doubt seen, observed, analyzed, reflected on the animal, but who have never been *seen seen* by the animal". They have, that is, "taken no account of the fact that what they call 'animal' could *look at* them, and *address* them from down there, from a wholly other origin". This being so, "they made of the animal a *theorem*, something seen and not seeing" (2008: 13, 14).

As we might have expected, Derrida sets out to do what so many philosophers before him so painstakingly have avoided: he describes what it's like to be looked at by an animal, a particular one at that: a cat, one that "looks at me naked", where "nudity gets stripped to bare necessity only in that frontal exhibition, in that face-to-face"; and "in these moments of nakedness, as regards the animal, everything can happen to me" (2008: 11, 12). What this scene actualizes is what Derrida terms the problem of "appellation" and of the response to a call. We should note that this is not the call-and-response exhibited in the interplay between a human and an animal, or between two animals, in the sense intended by Willett and discussed in Chapter 2. Instead, what Derrida has in mind is the eminently *ethical* meaning of responding to the other's call that is a leitmotif in Levinas, leading us to ask *who* issues the appellation (call) and who has what is required to respond.

So Derrida's cat (and his cat it is) can look at him, can see him in his nakedness, in the passivity Derrida associates with nudity. He speaks of the impropriety that can come of "finding oneself naked, one's sex exposed, stark naked before a cat that looks at you without moving, just to see"; "the impropriety of a certain animal nude before the other animal". Why this impropriety? Well, says Derrida, "it is as if I were ashamed, naked in front of this cat, but also ashamed for being ashamed"; "a shame ashamed of itself, at the same time specular, unjustifiable, and unavowable". But again, why be ashamed of this nudity before an animal, "naked as they are, or so it is thought, without the slightest consciousness of being so" (2008: 4)? That's precisely where we must look for the answer: animals' being naked without knowing it is what distinguishes them from humans, since to be without knowledge of their nudity is to be "without consciousness of good and evil". Accordingly, there is no nudity "in nature": "Because it *is* naked, without *existing* in nakedness, the animal neither feels nor sees itself naked" (2008: 5).

What is happening here? The scene where a man is naked before the gaze of a cat, also naked, might invite the assessment that they both being naked puts them on an equal footing, as it were, highlighting a similarity that gets lost once the man puts on his clothes. If anything, this in fact underlines their dissimilarity. Why? Because for the man being naked involves a choice, a conscious decision, a state of being he can undo any moment, or prolong as long as he wants, circumstances permitting. None of this applies to the cat. For the animal, being naked is a condition in the sense of something ineluctable, a given that defines their existence. Since the animal is naked without consciousness of being naked, it does not *relate* to its nakedness, and so does not feel comfortable in it, or uncomfortable; feeling the one rather than the other is not an issue for the animal, does not present options for it, which is to say that "modesty remains as foreign to it as does immodesty", and, by the same token, that it cannot be ashamed at its nakedness. Only for humans can nakedness be awkward or embarrassing, something that the person in question relates to, and is forced to relate to, as well as something that *other persons* will just as surely relate to, attesting to its significance, its lack of neutrality as it were, within both a wider cultural context and the specific social one in the given situation of being naked.

Derrida doesn't engage in comparison between being looked at naked by a cat and by another human person, a comparison that would have made discussion of Sartre's famous analysis of the (human) other's look as shame-eliciting, as stealing my freedom and transcending my transcendence all but inevitable. Sartre is never mentioned. That is just as well, because what Derrida is getting at is the question we know he associates with Levinas, namely whether the animal *responds*.

To answer that question we need to know how to distinguish a response from a reaction. Here Derrida invokes Lewis Carroll's work, in particular the scene where Alice, impatient and irritated, exclaims that whatever she says to her kittens, they *always* purr; and you can't talk with someone if they always say the same thing. So, in speaking to an animal, be it your cat, you will experience, time and time again, that "it doesn't reply, not really, not ever" (2008: 10). That is Alice's conclusion, arrived at not for lack of trying, as is Descartes'. Indeed, it is the received wisdom throughout modern Western philosophy. To hold that animals are incapable of responding is to hold that, for lack of actively (consciously) *relating to* whomever they encounter, they merely *react*. In merely reacting, they are not really agents, subjects.

Does this view survive closer scrutiny? Does it square with the experience that Derrida takes as his point of departure, that of being looked at by the cat? Would he feel ashamed at being looked at as naked by a cat, were it not for the cat's possession of what it takes to make him feel ashamed? How does *his* response reflect on the cat's powers in terms of agency? Insofar as he feels passivized, an object, when looked at by the cat, doesn't that go to show that the cat is indeed a subject not an object, active not passive?

Admittedly, this way of putting the question borrows not only from Sartre's analysis of the look but also from Martin Buber's distinction between an I-Thou relationship and an I-It relationship. And in fact, in a footnote Derrida does quote from Buber's work: "An animal's eyes have the power to speak a great language"; "the beginning of this cat's glance, lighting up under the touch of my glance, indisputably questioned me: 'Is it possible that you think of me? Do I really exist?'" (2008: 164, fn. 11). But instead of employing Buber's distinction, Derrida takes a detour, remarking that the animal question has been "left hanging" for about two centuries, in the course of which "we who call ourselves men or humans…have been involved in an unprecedented transformation", one that affects the experience of "what we continue to call imperturbably, as if there were nothing to it, the animal and/or animals" (2008: 24). What sort of transformation does he have in mind? He tells us that "in the course of the last two centuries the traditional forms of treatment of the animal have been turned upside down by the joint developments of zoological, ethological, biological, and genetic forms of *knowledge*, which remain inseparable from *techniques* of intervention *into* their object, from the transformation of the actual object, and from the milieu and world of their object, namely, the living animal" (2008: 25).

The transformation takes the form of an interventionist practice "in the service of a certain being and the putative human well-being of man", a practice whose cruelty "men do all they can in order to dissimulate or to hide from themselves", a violence which "some would compare to the worst case of genocide (there are also animal genocides: the number of species endangered because of man takes one's breath away)" (ibid.).

It is to Derrida's credit that he draws attention to these sinister facts about the plight of animals, not to mention their sheer disappearance by way of human-caused extinction, in today's world; there is nothing of the sort in his contemporary Habermas, as we saw above. But what bearing do these facts have on Derrida's attempt to answer the question about the moral status of animals, actualized in his experience of being looked at by his cat? With the megatrends seemingly locked to the trajectory of "industrial, mechanical, chemical, hormonal, and genetic violence" (ibid.), is a way forward to be found on the microlevel, that of one-to-one encounters such as that with his cat?

To be sure, there are attempts to fight the organized denial of the large-scale torture to which animals – mostly in the modus operandi of out of sight, out of mind – are subjected. Voices of protest are raised – "minority, weak, marginal voices", voices not confident of their discourse and its eventual enactment within the law, in the form, say, of a declaration of animal rights, thus appealing, Derrida says, "to what is still presented in such a problematic way as *animal rights*, in order to awaken us to our responsibilities and our obligations vis-à-vis the living in general" (ibid.). But rather than leading him to engage head-on with the notion of rights, Derrida turns at this juncture to compassion. In doing so he explicitly dismisses the thesis he sees maintained from Aristotle to Heidegger, from Descartes to Kant, Levinas, and Lacan, namely that animals are deprived of *logos*, proclaiming that Bentham's "Can they suffer?" is the question to ask. Why? Because in referring to suffering we have to do with a phenomenon possessing no less indubitable certainty than that claimed for the experience of thinking in Descartes' "I think, therefore I am". Bentham's question has exactly that kind of answer: "No one can seriously deny the suffering, fear, or panic, the terror or fright that can seize certain animals and that we humans can witness" (2008: 28). Indeed, the last two centuries that Derrida keeps referring to have been those of "an unequal struggle, a war (whose inequality could one day be reversed) waged between, on the one hand, those who violate not only animal life but even and also this sentiment of compassion, and, on the other hand, those who appeal for an irrefutable testimony to this pity" (ibid.).

This is a statement worth dwelling upon, not least because it is the clearest acknowledgment I have found in the text of the twin dimensions of the war Derrida talks about: *the simultaneity of the war waged against animals and the war waged against the sentiment of compassion*, the capacity in us that is the foundation for the struggle to end the war of humans against animal life. "Whoever is hard with himself earns the right to be hard with others", Adorno observed in his reflections on the socio-psychological profile of the Nazis leading the way in killing Jews (2005: 198). The violence performed against others has a twofold relationship to the violence conducted against oneself: the latter is both a presupposition for and a product of the former, in what over time amounts to a self-reinforcing dynamic. Conversely, you cannot bring the outer-directed violence to an end without ending the self-directed one. True, Adorno wasn't thinking about the violence against animals in making his observation, but the violence against other humans. Yet he would certainly not want to confine the dynamic at work to the relationship between humans. *Pace* Descartes, the reality of suffering is just as indisputable in animals as in humans, their fight against pain and their agony in being subjected to it just as undeniable as in us; accordingly, the pain humans willfully inflict on animals is just as much a moral issue as that inflicted on fellow humans. The common denominator is the vulnerability in being exposed to suffering, to pain, in both humans and animals; and the capacity that resides in humans for protesting against that suffering, in animals as well as in humans. To render this capacity a one-species-only one in its orientation and commitment, meaning "fellow humans only", is the work of culture and society, passed on from one generation to the next by patterns of socialization whereby animals are excluded from the moral universe; a work against our own capacity for compassion, and whose intention to create indifference when confronted with *animals'* suffering risks producing it towards that of other humans as well.

It is no coincidence that I draw attention to Adorno. If Derrida had taken into account his *Negative Dialectics* and not limited himself, erratically, to his book on Beethoven, he would have been able to elaborate more fully on the crucial role of affectivity in making the case for a notion of solidarity that incorporates animals alongside humans. Adorno speaks about "the impulse of solidarity with what Brecht called 'tormentable bodies'", contrasting the feeling of solidarity with the "rationalizing" of morality he identifies in Kant's formulation of a moral law of equality, a law Kant of course held to apply only to humans. Solidarity in the Brecht-inspired sense is no abstract principle but rather a somatic impulse or "spontaneously stirring impatience" (1973: 286). Indeed, Adorno locates the core of morality in something for which "Kantian ethics – which accords affection, not respect, to animals – can muster only disdain: to try to live so that one may believe himself to have been a good animal" (1973: 299). Christoph Menke's comment is worth quoting:

> The action stemming from a feeling of solidarity is the action of a "good animal" – of a subject that does not separate itself from its "forces" or "impulses" for the sake of following the law and in order to make itself freed from them but rather whose freedom, indeed, whose very strength, consists of allowing its forces or impulses to express themselves. Only

> in this way, in "harmony", even in "reconciliation", with himself, with his powers and impulses, can man be good to others. As Adorno puts it: "Men are human only where they do not act, let alone posit themselves, as persons" (1973: 277).
>
> (Menke 2004: 320)

Having Kant as its critical target, this emphasis on affectivity as linked to impulses in a somatic sense no doubt has a Nietzschean ring to it in promoting the expression of precisely those aspects of humans' "inner nature" that, owing to their animal-likeness, Kant endeavoured to suppress, regarding them as coming in the way of what characterizes humans as moral agents *and* as appropriate moral addressees, namely the rationality that in both instances founds their status as *persons*. This is what motives Adorno's beg-to-differ statement that "Men are human only where they do not act, let alone posit themselves, as persons". But whereas Nietzsche's critique of the rationalist Kantian morality of equality rests on his wish to expose the ambition for equality as deceitful and as arising from *ressentiment*, as do Christian and compassion-based ethics as well, according to Nietzsche, for Adorno the somatic impulse giving rise to solidarity is one that should *not* be disqualified, neither for Kantian reasons nor for Nietzschean ones. Why? Because for Adorno "the physical moment tells our knowledge that suffering ought not to be, that things should be different. Woe speaks: 'Go'" (1973: 203).

The view that the existence of suffering in itself dictates this "go", which is to say this "no", that it should not be, and that it should be brought to an end, is known as *moral realism*. It holds that the phenomenon as such (say, suffering) is non-neutral; it contains something of normative import, simply in virtue of being the phenomenon it is. The normativity that thus resides in the phenomenon, that is met upon there, as opposed to being projected onto it, attributed to it, is such as to issue a command so as to *respond to it in certain manner* – that of trying to stop the suffering, recognizing it as something that should not be. That it should not be is – if we need use the word – the meaning of (intrinsic to) the phenomenon. To fail to respond in this manner, with this objective, is to fail to identify the phenomenon correctly, for what it is and *therefore* demands in terms of response.

True, Adorno doesn't put it in these terms, but it doesn't require his using the label moral realism to recognize it as his position, as Jay Bernstein has argued in great detail (2001). Note that in talking about how the phenomenon at hand "issues a command" I deliberately allude to Levinas and what he says about the face as the site of the appeal "not to kill": that is the meaning (again, if that word is apt) that resides in the face, in keeping with moral realism as just set out. To illustrate with a formulation of Adorno's: "The lines 'No man should be tortured; there should be no concentration camps' are true as an impulse, as a reaction to the news that torture is going on somewhere. They must not be rationalized; as an abstract principle they would fall promptly into the bad infinities of derivation and validity" (1973: 285f.).

We saw that in Derrida's reading of Levinas, "Thou shalt not kill!" does not forbid humans to kill an animal; it forbids only the murder of the face, and the animal fails to qualify as a face according to Derrida's interpretation of Levinas. He considers Levinas to remain "profoundly Heideggerian" in holding that, inasmuch as the animal is incapable of being the victim of murder, it doesn't die – only humans do (2008: 110). Now, if one can put the animal to death without "killing" it or murdering it, this is because "the animal remains foreign to everything that defines sanctity, the separation and thus the ethics of the person as face" (2008: 111).

This is important for the issue central to my overall discussion, namely how to conceive of the relationship between being a moral agent (subject) and being a moral addressee. In my understanding, to be an addressee is to be the object of responsibility on the part of some agent. I follow Levinas in holding such responsibility to be prior to intentionality, deliberation, and calculation, that is, prior to choice, prior to contract, and in keeping with moral realism I hold responsibility to take the form of responding to an imperative issued by the other, exemplarily, by perceiving that other's vulnerability to suffering, eliciting the "go" we saw Adorno speak about above. To repeat, the promise I see in the ethics of Levinas as distinct from, say, Kant, is that it recognizes that at its core the relationship between agent and addressee is one of dis- or asymmetry: the addressee does not have to be an agent. Ethics is not about reciprocity, does not rest upon (actual or potential) reciprocity as its rationale or precondition or goal; it is fundamentally about responsibility for the other.

Given this view of the structure of the ethical relationship, let's examine what Derrida says about responsibility when interrogating Levinas' position on animals. What does he mean when stating that "the animal remains foreign to everything that defines sanctity"?

What Derrida is getting at is that in Levinas, responsibility is linked to "Here I am"; it implies self-representation in an autotelic, auto-biographical sense, "exposing oneself before the law", which in turn implies the possibility of "responding", of "answering for oneself in the response to the appeal or command of the other" (2008: 111). This is the crux of the matter: "The animal, according to Levinas, seems deprived of all possibility, in fact, of all power of saying 'Here I am' and of responding, hence of all responsibility. It is nobody, certainly not a 'person', a word that Levinas always kept intact. That is why the animal doesn't die. [...] Death is not for Levinas, in the first instance, a passage from being to nothingness, an annihilation, but, as he often says, the moment when the other no longer responds" (ibid.). Whereas to state in the case of the dead face that "he doesn't respond" means "he is no longer responding", to state it in the animal's case means "it has never responded" and "it will never respond" (2008: 112). Why? Because the animal will never have been able to respond. We thus have to do with a double refusal, a twofold denial: not only is the animal "deprived of the power and the right to respond, and therefore responsibility"; it is also deprived of nonresponse, of "the right of nonresponse that is accorded to the human face by means of secrecy or in death" (ibid.). Given this double refusal, might there still be a place for the animal-other in Levinas, say, as what he calls "the third person", taken as the first appeal of justice? "But no", finds Derrida: "When Levinas reflects on the other of the other who is not simply a fellow [*semblable*] and who brings the question of justice to the fore, that nonfellow remains human, a brother and not another other, not an other other than the human" (ibid.; see Derrida 1999: 32ff.).

To sum up, then, on Derrida's reading the animal comes up short ethically speaking in Levinas no matter in what role or condition you consider it, be it as the "Here I am", be it as the other to respond to, to be addressed by in the form of "Thou shalt not kill", be it as "the third" who raises the question of (necessity of) justice; be it as living, be it as no longer living. In its very originality, concludes Derrida, what remains is a "profound anthropocentrism and humanism" (2008: 113). Rather than privileging the question and the request of the animal, as one might have expected from an ethics devoted to "a thinking of the other", what one finds in Levinas is exactly what one finds in Descartes and Kant before him: the same disavowal and foreclosure directed at the animal, "just as powerful when they don't speak of it or when they speak of it in order to deny to the *animot* everything they attribute to the human" (ibid.). Quoting once again from Levinas' essay on Bobby – "At the supreme hour of its institution, with neither ethics nor *logos*, the dog will attest to the dignity of its person. That is what the friend of man means" – Derrida comments that it (the dog) "can witness to us only for us, being too other to be our brother or neighbor, not enough other to be the wholly other", leading him to contend, by way of summing up: "The animal remains for Levinas what it will have been for the whole Cartesian-type tradition: a machine that doesn't speak, that doesn't have access to sense, that can at best imitate 'signifiers without a signified'...a sort of monkey with monkey talk, precisely what the Nazis sought to reduce their Jewish prisoners to", the latter being a reference to Levinas' statement "How can we deliver a message about our humanity which, from behind the bars of quotation marks, will come across as anything other than monkey talk?" (2008: 117, 118).

I had to rub my eyes when I first read the sentence where Derrida charges Levinas with replicating in his account of animals "precisely what the Nazis sought to reduce their Jewish prisoners to". Really?

A couple of pages before he states this, Derrida refers to "The Paradox of Morality", the interview in 1986 where Levinas is asked the question "Is there something distinctive about the human face which, for example, sets it apart from that of an animal?" However, Derrida does not quote the following part of Levinas' answer, by far the most substantial one:

> One cannot entirely refuse the face of an animal. It is via the face that one understands, for example, a dog. Yet the priority here is not found in the animal, but in the human face. We understand the animal, the

> face of an animal, in accordance with *Dasein*. The phenomenon of the face is not in its purest form in the dog. In the dog, in the animal, there are other phenomena. For example, the force of nature is pure vitality. It is more this which characterizes the dog. But it also has a face. There are these two strange things in the face: its extreme frailty – the fact of being without means and, on the other hand, there is authority. It is as if God spoke through the face.
>
> (Levinas 1988a: 169)

Later in the interview, Levinas adds the following, quoted as we saw above by Derrida: "It is clear that, without considering animals as human beings, the ethical extends to all living beings. We do not want to make an animal suffer needlessly and so on. But the prototype of this is human ethics. Vegetarianism, for example, arises from the transference to animals of the idea of suffering. The animal suffers. It is because we, as humans, know what suffering is that we can have this obligation" (1988: 172).

Do we meet here a philosopher who "denies to the *animot* everything he attributes to the human", to recap one of Derrida's claims? Is Derrida open to what Levinas is saying? Or does he find what he is looking for, and nothing else?

Whatever the reasons for Derrida being so demonstrably uncharitable to Levinas in this text, we are still searching for his position on the two major issues I raised above, the one concerning the link between responsibility and response, the other concerning the nature of ethical relationship as profoundly asymmetrical. Given his criticisms of Levinas, where does Derrida turn to sketch an alternative view?

He turns to Jacques Lacan, the seminal figure in French psychoanalysis in the last part of the twentieth century. To Lacan, the animal has neither an unconscious nor language; hence the animal remains within the imaginary, unable to accede to the symbolic. Lacanian parlance aside, what this means is that Lacan subscribes to the hegemonic view according to which animals are only capable of *reactions* to stimuli, not *responses* to questions. While bees may appear to "respond" to a "message", they do not respond like humans would do, they merely react in the sense of obeying a fixed program, one they cannot deviate from or transcend. Lacan goes on to found responsibility on the distinction between reaction and response – the distinction, Derrida tells us, that he "find[s] so problematic" (2008: 125). Indeed, he wants to cast doubt on "the purity and indivisibility of a line between reaction and response, and especially the possibility of tracing such a line between the human *in general* and the animal *in general*" (2008: 126). In doing so, he is acutely aware that he will be seen as casting doubt on "all responsibility, all ethics, every decision, etc.", a risk he is more than willing to take, since casting doubt on one's own being-ethical and being responsible should be regarded as the very essence of ethics.

Turning to Lacan's text "The Subversion of the Subject and the Dialectic of Desire in the Freudian Unconscious", not the distinction between reaction and response but the difference between pretence and deception proves to be at the heart of Lacan's account of the animal. Whereas the animal is quite capable of pretence (warrior, predatory, or seductive suit, pursuit or persecution), it is incapable of "the deception of speech within the order of the signifier and of Truth" (Derrida 2008: 127f.), that is to say, deception inasmuch as it presupposes speech, occurs within speech and between parties in command of speech. In particular, this means that the animal would not properly know how to lie, even if we grant that it knows how to pretend. Or, to be more accurate, Lacan allows for the first degree of pretence in the animal (pretence without pretense of pretence); what it does not do is pretend to pretend, since to do so it would have to break out of the presymbolic. Agreeing with Heidegger that the animal is "a living creature that is only living", that for lack of a capacity to *relate to* its own mortality – indeed, to relate to anything at all "as such", as the experience or object it is – it does not die, Lacan stresses the importance of distinguishing deception as practiced in speech, within the order of the signifier, and the kind of pretence to be found in physical combat or sexual display, or when animals are being hunted (discussed above), where they show their ability to put their pursuers off the scent by making a false start. Yet practices such as vouching for one's word, or making a promise, are precisely what animals are incapable of, according to Lacan, presupposing as they do the gift of speech. That animals do not give their word has its corollary in the fact that a human wouldn't give its word to the animal; thinking to do so would be a case of projection or anthropomorphic transference, as if a human's promising something to an animal would make a difference to it.

Why is this reasoning on the part of Lacan important to Derrida? And what light does it shed on the issues most central to his critique of Levinas, concerning responsibility and asymmetry?

To get to that, we need to note that Derrida questions Lacan's claim that the animal *in general* is incapable of pretending pretence. For lack of offering ethological research to back up his claim, the refusal to the animal of the pretence of pretence is nothing but "a simple dogma" (2008: 133). Pressing further, Derrida finds that the distinction between lie and pretence becomes precarious, likewise that between Speech and Truth (in Lacan's sense). Hence Lacan is not justified in asserting that the animal *in general* does not cover its tracks. Yet what is at issue here, for Derrida, is not whether Lacan is successful in refusing animals some particular power or capacity (speech, lying, pretence of pretence, covering of tracks), or whether he has the right to embark on the project of animal-directed refusal in the first place, following in the footsteps of so many Western thinkers before him. More important, says Derrida, is "asking whether what calls itself human has the right rigorously to attribute to man, which means therefore to attribute to himself, what he refuses the animal, and whether he can ever possess the *pure, rigorous, indivisible* concept, as such, of that attribution" (2008: 135). Even if we grant that the animal is incapable of covering its tracks, something Derrida is unwilling to do, the truly vital question to ask is "by what right could one concede that power to the human, to the subject of the signifier" (ibid.)?

Unsurprisingly, this question leads Derrida to zoom in on Lacan's constant appeal to the Cartesian *cogito*, the "I think" that is intended to allow for absolute certainty, a certainty that is, and must be, established *in foro interno*, in the I's relationship with itself as distinct from with something – or someone, co-human or animal – met upon in the outside world. But of course, for Lacan things are not that simple, not that transparent. The Cartesian *cogito* has been destabilized at least twice over since Descartes theorized it as a unique "Archimedean point" (Hannah Arendt); first by Hegel, most famously in his Master/Slave dialectic in *The Phenomenology of Mind*, and then by Freud's psychoanalytic theory, especially in the analysis of the Oedipus complex. In Lacan's terminology, "the becoming-subject of the ego passes by way of the signifier, Speech, Truth, etc., that is to say, by losing its immediate transparency, consciousness as consciousness of the self identical to the self". This dynamics, says Derrida, "ends only in an apparent paradox: the subject is confirmed in the eminence of its power by being subverted and brought back to its own lack, meaning that animality is on the side of the conscious ego, whereas the humanity of the human subject is on the side of the unconscious, the law of the signifier, Speech, the pretended pretense, etc." (2008: 137f.). Contrary to the certainty arrived at by way of transparency associated with the Cartesian cogito, where the thinking I in self-sufficient manner establishes itself as the indubitable ground from which all epistemic certainties spring, Lacan exposes the lie and the deceit "in the very heart of the cogito" (ibid.), rendering the transparency-cum-certainty Descartes had invested it with profoundly deceitful and so the whole undertaking on the part of the ego an exercise irredeemably liable to self-deception. Accordingly, the question of the I – "What am I?" or "Who am I?" – shows up a lack, says Lacan; it exposes "what the subject lacks in order to think himself exhausted by his cogito, namely, that which is unthinkable for him" (2008: 139). He (sic), Lacan continues, "lacks everything needed to know the answer" to the question of where his being, which "appears in some way defective in the sea of proper nouns", originates (ibid.).

Where does all of this leave the animal? Has the animal question been lost along the way? If so, is its being lost the very point Derrida is intent on making? A point that can be made over and over again, forming as it were a red thread in the Western canon?

If Derrida demonstrates that the animal question is lost, that it slips away, again and again, in all the thinkers he interrogates, then his whole exercise amounts to a series of negative findings and conclusions, a lost opportunity. Is getting to the bottom of the question of the animal tantamount to witnessing how it is effaced in each and every endeavour to vindicate it? Or is the truth of the matter the banal one that Derrida has picked the wrong cast of philosophers, so that a different selection would have yielded a more positive result?

As far as I am concerned, the selection is excellent. And by choosing the experience of being looked at by his cat as his point of departure, Derrida seems positioned to grasp the peculiar nature of the other as animal, the animal as the encountered Other. So what happened?

What happened is that Derrida's initial openness by way of being exposed qua naked to the gaze of the cat is effectively closed once he starts asking questions about it. Intellectualization sets in, tightening its suffocating grip on the subject – the cat as subject, as the one looking – ever after. The opening encounter with the cat, taking Derrida the human party to it aback, causing him to lose his privileged point of view, to abandon it *qua* privileged in the sense of superior, if only for a moment, this access-opening moment is not allowed to play out; instead it yields to the human-philosopher's retreat to his abode, that of the safety of considering-at-a-distance what the encounter amounts to, and the consequences that would follow, some of them presumably ethical, bearing on how to respond to the cat in the Levinasian sense of responsibility.

What got lost? Or who? Derrida's analysis lost sight of the animal it set out to engage with because Derrida all-too-quickly transformed it into another question, a question for another subject than the cat – namely, the question about how he as a human can possibly respond to an animal, leading to a self-interrogation – the human self's interrogation of itself – that, eminently Cartesian as it is, denies participation from the animal. While on the one hand the question of the animal's ability to respond to the human remains tied to the nagging suspicion that it "merely reacts", one the other hand the human's response to the animal is transposed from its directedness at the animal being encountered to an inspection of the human's interiority in terms of possessing or not possessing the capacities it would arguably take to respond, never to leave that introspective gaze – not the animal's gaze at the human, but the human's at himself, *gut Cartesisch* – after having adopted it.

It is doubly ironic that Derrida should lose sight of his cat like this, considering that he spends so much time on Levinas' contribution to the animal question. Even if one grants that Levinas' answer to the question whether he considers an animal to be a face is far from satisfactory (as Derrida asserts), it is puzzling to witness how negligent he is about the one intuition that carries Levinasian ethics, namely that the response-cum-responsibility to/for the Other is prior to ontology and knowledge, to the question of being and the quest for knowledge. That Derrida remains committed to conceiving the question about the animal within the orbit of precisely the twin orders that Levinas breaks with in order to establish response-responsibility as pre-cognitive and pre-intentional, is attested to again and again, perhaps nowhere more conspicuously than in his critique of Lacan, where what fuels his asking what gives man (sic) the right to attribute to himself what he refuses the animal is, ultimately, "whether he can ever possess the *pure, rigorous, indivisible* concept as such, that of attribution" (2008: 135). When everything is made to hang on issues of concepts and capacities, of what justifies whatever is being attributed cognitively by humans to animals, and by humans to themselves, the result is that the animal disappears.

Was not the whole point of Derrida's effort to avoid, to subvert, to reverse this very disappearance? If what he finds – be it in Descartes, in Kant, in Heidegger, in Levinas, in Lacan – is so many variations of the same refusal to let the animal "in", the same insistence that it must remain outside of the moral universe, then what should he do?

In my view, there are two occasions in particular where Derrida could have taken a more constructive turn, instead of getting so entangled in the targets of his interrogation as to reproduce rather than subvert the denial of animals that he identifies.

First, why doesn't he linger longer with the cat's point of view? Why is he so impatient to move on, beyond it – only to return to his human-specific one? Why does he disqualify every trait of kinship-likeness with the cat that he considers, or more to the point: that he initially experiences – undergoes, is subject to – in the Levinasian sense of exposure to the Other as prior to intentionality and cognition; disqualify them by immediately shifting emphasis to the traits in question *being more about dissimilarity*, whereby dissimilarity is always taken by him, as by those he criticizes, to speak *against* the inclusion of the animal, the solidarity with it that Adorno insists on? If animals are partly similar to humans, partly dissimilar, why regard the latter as always more intriguing, more important and exhibiting the crux of the matter, ethically speaking?

True enough, says Derrida, the cat is naked like he is, but its nudity by way of not wearing clothes carries a wholly different meaning than does his, being a decision on his part, never on the cat's. True enough, the cat looks at him, but the shame it elicits on his part shifts the attention from the cat, the cat both as subject and as object (other), that is, by way of retreat to the human-being-ashamed-as-only-humans-are, introspecting there for its explanation and justification in a way that has everything to with the perspective of the human-in-the-world, nothing with the cat's. And true enough, the human and the cat are both vulnerable to suffering, to the reality of pain; but once that capacity is thematized, the cat's lack of sense of temporality, ultimately of perceiving pain as at bottom pointing to the possibility of death, to an active relating to one's own death, highlights how differently the similarity in exposedness to pain plays out in the lives of humans and cats. And so it goes, on and on.

Second, why follow Lacan in putting such emphasis on the animal's "profound innocence", its being incapable of the "signifier" peculiar to the symbolic order, to speech – its being incapable of lying and deceit, of pretended pretence, linked, we saw Derrida find, to the "most traditional" notion of "the cruel innocence of a living creature to whom evil is foreign, living anterior to the difference between good and evil" (2008: 130)? Why is this crucial? Put aside for now Derrida's claim that Lacan lacks the evidence to prove that the animal "in general" is incapable of pretending pretence and of covering its tracks, that he merely repeats a simple dogma. The question is: what justifies accessing the lives of animals through the hermeneutics of suspicion? Why highlight the phenomenon of deceit? Why do so to the detriment of appreciating animals' capacities for care, trust, and loyalty? In trying to make sense of the cat looking at him, is its capacity to deceive and cover its tracks really what commands attention? And why is animals' capacity or lack of such for evil so important? What is meant: evil to other animals, or to humans? Is this capacity required to qualify as a moral subject? Or as a moral addressee? Or both?

To ask such questions is to recognize how poorly they fit with Derrida's approach. He starts out admitting his cat the role of subject, only to shift the whole analysis to the human-all-too-human embarrassment of being looked at, whereby this response to the cat's look revolves entirely around the question of what difference it makes for *Derrida* to be objectified by a cat's look as distinct from a human's. The obvious corollary is not dwelt upon: what difference it makes to the cat to look at a human as distinct from another animal.

To return to the capacity for evil: many would agree that the capacity for evil, understood as the alternative choice to good, is crucial to being a moral agent, but not to being a moral addressee. Although Levinas doesn't put it like this, I take him to endorse the view. This goes to confirm the asymmetry I have talked about between agent and addressee, asymmetry not only in terms of difference in power but also in terms of difference in capacities. Brought to bear on what we are exploring here: Whether the animal is capable of deceit and of evil makes no difference with regard to its status a moral addressee, only to its status as a moral agent. Again, the point of the asymmetry is (among other things) that you don't have to be an agent in order to be an addressee.

That Derrida does not make this distinction has consequences for his case against Levinas. What gets lost is his ability to address the asymmetry that is the signature of Levinas' ethics. In saying this I do not take Derrida to task for *not agreeing* with the asymmetry and its implications; I take him to task for not recognizing its importance and, as part of that, the way it shows Levinas to break radically with rival conceptions of the moral relationship.

Of course, nobody knows the importance of asymmetry a la Levinas better than Derrida – one can say that his book (based on his speech at the funeral) *Adieu to Emmanuel Levinas* is one long meditation on it. But that does not change what I just said. The point I am making goes to the heart of what Derrida considers the core of Levinas' teaching: the question whether the animal has a face. Why is that important? Because for lack of a face the animal would not be able to issue the command "Thou shalt not kill". As we saw, according to Derrida there is in Levinas no such command forthcoming from the animal. And since only the face forbids murder, understood as the murder of the face, denying the animal a face means that "putting to death or sacrificing the animal, exploiting it to death – none of those, within this logic, in fact constitutes murder" (2008: 110).

Leave aside for now that I have made it clear that I don't agree with Derrida that Levinas denies the animal a face; his full answer to the question tells a different story, as we saw above. What matters here is the connection Derrida attributes to Levinas when stating that "where there is evil there is face. What remains faceless is pure indifference to good and evil" (2008: 110). I take the implication of the statement to be that for lack of being a face, anything goes, with impunity; no suffering brought upon what is faceless would qualify as morally wrong, let alone as evil. But to hold that "where there is evil there is face" has a corollary: where there is face there is evil. This corollary, not entertained by Derrida, helps us realize that the distinction between agent and addressee is missing from his discussion of the significance of evil. For lack of making this distinction, the statement "What remains faceless is pure indifference to good and evil" is ambiguous: what is meant – indifference to *doing* good and evil, or indifference to *suffering* it?

The issue of evil is a good opportunity to sort out the difference between agent and addressee, one squandered by Derrida. I take Derrida and Levinas to agree that ability to suffer is the pivotal criterion for moral consideration, for being an addressee. Does everything (everyone) that can suffer have a face? Probably not, for the two are not identical: there are more creatures in nature able to suffer, in virtue of sentience, than there are creatures about whom we will say that they have a face. Moreover, it is probably easier to agree about ability to suffer than to agree about qualifying as face. This goes to show that Derrida is onto something important in pressing Levinas concerning his readiness to "attribute" a face to the animal, or more to the point, to some animals but not to others. The deeper question, though, is to do with why being a face should be considered necessary to qualify as addressee, as an apt "other" for the agent's responsibility. If there is no one-to-one-correspondence between suffering and face, such that the capacity for suffering need not be tied to "being a face", need not depend on it to qualify morally speaking, why is it such a scandal for Derrida that (on his reading, mind you!) Levinas is hesitant to attribute a face to all animals, and *a fortiori* to all sentient beings? This is a scandal only because Derrida is committed to being-a-face having a pivotal ethical significance. But isn't that exactly the notion he should question? What goes to show that Derrida is not implicated in the same scandal he pins on Levinas? If Derrida defends the position alluded to, that there may be faceless beings able to suffer and that it is their ability to suffer and not their not-being-face that ethically counts, why doesn't he make a sustained argument to that effect?

This is a complicated issue, for both thinkers. For one thing, and notwithstanding Derrida's pretending not to know, Levinas *does* say "It is clear that, without considering animals as human beings, the ethical extends to all living beings" (1988: 172); that is to say, it extends not only to more living beings than humans, meaning to animals too; but to more non-human creatures than animals as well, since not all non-human beings are animals. But what would it mean, in practice, to extend moral status and, correlatively, moral responsibility towards *all living beings*, based on their sentience-based capacity for suffering, and thus their vulnerability to unnecessary pain so as to qualify as harm, as someone's doing the being (creature) in question wrong?

Of course, to explain what it would mean to hold that all living beings are part of the moral universe would require the defence of some version of eco- or biocentrism. Neither Levinas nor Derrida as much as hint at embarking on such a project, associated as it is with, say, the deep ecology of Arne Næss (1989). Instead they both have their hands full with humans and animals, never raising such questions as whether cutting down a tree is wrong, a wrong done to the tree, at that. This being so, let us return to the less radical view that *animals* are part of the moral universe. In affirming that view, what precisely do we mean to admit to animals, morally speaking?

I take it that the answer is: that animals are entitled to moral consideration. In other words, they qualify as moral addressees. But what about animals as moral agents? My position is clear: animals do not have to be agents in order to be addressees; nobody does. I call that the asymmetric structure of the ethical relationship, and I take Levinas to embrace it.

But this only takes us so far. We need to be more precise. It is not only that animals don't *have to* be themselves moral agents in order to qualify as addressees, their capacity for suffering sufficing for the latter. It is that they *aren't* agents, not in the full-blown sense of *moral* agency I am committed to (recall that I observed as much in my above discussion about Bobby). Why is that? Essentially, because animals (and here I do speak generally, generically) lack the capacities needed for responsibility and so, as part of that, for acting in a morally blameworthy manner, so that critiquing their action along such lines would be misplaced and unfair. It would be to commit animals to partaking in a sort of practice that is uniquely human, that is, the practice where moral agents critique and sanction each other for failing, in the concrete instance, for living up to the requirements that go with enjoying the status of being such agents – it is a gift that comes with strings attached, to do with being eligible for moral criticism. In keeping with this view, animals are not moral agents because they are incapable of evildoing; they are incapable of killing in the morally relevant sense of committing murder. Only humans commit murder. Only humans, that is to say, are capable of *immorality*. That is an important reason why only humans are moral agents.

This is the point I alluded to above, the point Derrida had a perfect chance to make when faulting Levinas for (allegedly) not being prepared to acknowledge that a human's killing of an animal is morally wrong in the sense of being murder; animals can be murdered just as humans can. Yet in terms of morality, of participating in a comprehensive moral universe not restricted to humans only, the question to ask is not only Derrida's "Can an animal be murdered?"; it is also "Can an animal murder?" To ask the second question is to interrogate what is required to be a moral agent, given that only such ones can murder, that is, can do something to somebody that is a wrong in a sense implying blameworthiness. It takes addressees – proper ones – to be an agent; and being an agent towards the proper addressees means being able to wrong them. In this sense, the possibility of immorality is the condition of morality. And the possibility of immorality depends on the mortality of the addressee (the other, all proper others), mortality understood as a dimension of the latter's vulnerability, susceptibility to suffering (see Vetlesen 1994, 2009). In this way, mortality is the *sine qua non* of morality.

I do not claim that this is the reasoning I find in Levinas and/or in Derrida. It is my own, and its relevance for my overall argument will become clearer when we proceed to the topic of extinction, in the next chapter – extinction being a phenomenon that forces us to appreciate head-on the connection I propose between morality and mortality.

IV

Having completed my discussion of Derrida, I turn now to a separate discussion of Levinas. I begin by looking at how Levinas understands the connection between responsibility and the choice involved in evildoing. Consider this passage from his essay "Humanism and An-archy":

> The insurmountable ambiguity of Evil is its essence. Seductive and facile evil is perhaps incapable of breaking the passivity of the pre-liminary, pre-historical subjection, annihilating the hither side, repudiating what the subject has never contracted. Evil shows itself to be sin, that is, a responsibility, despite itself, for the refusal of responsibilities. It is neither alongside of nor in front of the Good, but in the second place, beneath, lower than, the Good. *The being that perseveres in being, egoism or Evil, thus outlines the dimension of baseness itself, and the birth of hierarchy.* Here begins the axiological bipolarity. But Evil claims to be the contemporary, the equal, the twin, of the Good. This is an irrefutable lie, a Luciferian lie. It is the very egoism of the ego that posits itself as its own origin, an uncreated, sovereign principle, a prince.
>
> (1993a: 137; my italics)

In no uncertain terms, Levinas maintains that his ethics does not affirm the equiprimordiality of good and evil, that is, the theodic notion that governs so much of Western mentality to this day. And he adds, following the passage just quoted: "The ego, brought back to itself, responsible despite itself, abrogates the egoism of the *conatus* and introduces a meaning into being. [...] But the pre-original responsibility for the other is not measured by being, is not preceded by a decision, and death cannot reduce it to absurdity. Yet we can have responsibilities for which we cannot not consent to death. It is despite myself that the other concerns me" (1993a: 138).

I cite these passages partly because they are conspicuously ignored in Derrida's discussion of Levinas, and partly because they alert us to something that I consider absolutely crucial, namely Levinas' view of the "egoism of the *conatus*", casting its concern with persevering in being as tantamount to "egoism or Evil", as outlining "the dimension of baseness itself". The "shock" of responsibility, its being unasked-for, non-optional, in the sense we are getting at here means for the self (ego) to be "always discovered, exposed, to turn the other cheek, which is the expiation for violence undergone by the fault of another" (1993a: 133, note 11). Of course, the ring of Christian thought is not to be overlooked; "expiation" means "pay the penalty of, make amends for (sin)", according to the Oxford Dictionary that sits on my desk. But what is this "violence", by and against whom is it committed? And what justifies the assertion of affinity, if not one-to-one correspondence, between the "egoism of the conatus" on the one hand and "evil" on the other?

So far we may have got the impression that according to Levinas there is no I (or self, or ego) prior to the encounter with the other. Yet the first part of Levinas' first major work, *Totality and Infinity*, tells a different story. Here Levinas develops a rich phenomenology of enjoyment and need, painting a condition where the self bathes in the elements, hungry for the warmth of the sun, the fertility of the soil, the coolness of water. There is "nature" everywhere, generously providing what the self needs not only to keep on living, but to feel joy and vitality in doing so. Only there are no others. Not yet. So, somewhat surprisingly, there is a pre-moral and pre-social mode of existence in Levinas, such as to invite comparison with Hobbes. Yet, unsurprisingly, the two pictures drawn are very divergent. In Levinas' depiction, existence is not harsh, it is enjoyment. I do not start out as a foreigner in the world, thrown – without warmth, without security and trust – into a life of scarcity, a life dictated by that severe and cruel master, Nature. There is instead immediate belongingness in the world; it is mine, I am at one with it, and at one with myself in being so. There is dwelling; I have a home, I am surrounded by things. They are near and dear to me by way of my making use of them, mending them, needing them. I literally *live from* the elements of my existence in the world; I live from "good soup", air, light, spectacles, work, ideas, sleep. These are not so many things perceived first as objects of representations. Nor are they tools or implements in the sense known from Heidegger's *Being and Time*; their utility neither defines them nor exhausts their existence. Instead they are always in a primordial way "objects of enjoyment, presenting themselves to taste, already adorned, embellished"

(1991a: 110). Thus things are always more than the strictly necessary; they make up “the grace of life”. Life is *love of life*; the reality of life is “already on the level of happiness”, a relation with contents that are “not my being but more dear than my being: thinking, eating, sleeping, reading, working, warming oneself in the sun” (1991a: 112). True, life is not pure happiness, pure enjoyment; we are also beings having needs. But needs do not spoil happiness, do not frustrate enjoyment. It is the other way around: “What we live from does not enslave us; we enjoy it.” The human beings “thrives on his [sic] needs; he is happy for his needs”. Before anything else, therefore, there is complacency and pleasure: “Living from...is the dependency that turns into sovereignty, into happiness – essentially egoist” (1991a: 114).

What matters at this stage of existence is this: “In enjoyment I am absolutely for myself. Egoist without reference to the Other. I am alone without solitude, innocently egoist and alone. Not against the others, not ‘as for me’, but entirely deaf to the Other, outside of all communication and all refusal to communicate” (1991a: 134).

The I (or self, or ego) as portrayed here is thus perfectly self-sufficient, preoccupied with enjoying things in the very act of needing them, using them. It is an I immersed in the primordial positivity of enjoyment, a positivity opposed to nothing and therefore sufficient to itself from the first. This, then, is the I as it exists before the entry into its existence of the other. So the question is: How does this I receive the other? Does receiving the other require a *different* I from the one portrayed here? Is a *shift* from a premoral mode of existence to a moral (morally charged) one involved, from primordial egoism to genuine other-directedness?

Now the “egoist” painted by Levinas is no fighter in a Hobbesian war of all against all. It is rather that the other does not (yet) matter, and that – coming as an intruder at night – all of a sudden he or she does. The other passes from irrelevance to primacy.

In *Totality and Infinity*, Levinas tells us that “The surpassing of phenomenal or inward existence does not consist in receiving the recognition of the Other, but in offering him one’s being. To be in oneself is to express oneself, that is, already to serve the Other. The ground of expression is goodness” (1991a: 183). So the Other is received by the I in the mode of goodness. Levinas also speaks of gentleness: “The Other precisely *reveals* himself in his alterity not in a shock negating the I, but as the primordial phenomenon of gentleness”. He adds: “The welcoming of the face is peaceable from the first, for it answers to the unquenchable peacable Desire for Infinity”. This “peacable welcome is produced primordially in the gentleness of the feminine face, in which the separated being can recollect itself, because of which it inhabits, and in its dwelling accomplishes separation” (1991a: 151).

Here, then, the other is welcomed in at atmosphere informed by the goodness and gentleness of the welcoming I, nourished by its prior state of happiness fostered by the generosity of the elements, providing for every need. If there is at all a transition, a shift, undergone by the I who welcomes the other here, it appears to be perfectly undramatic one. There is no metamorphosis; none seems required.

But the drama that is absent in Levinas' first major ethical work is all the more at the forefront in the second one, *Otherwise than Being or Beyond Essence*, so much so as to recall another book title, dealing with a different topic: *This Changes Everything* (Naomi Klein). Now the entry of the Other is said to turn the tables completely, disturbing and disrupting everything that made for calm, happiness, and peace in the existence of the self, in its entire world. Why? Because in responding to the Other a responsibility is posited for – not by, as in choosing and deciding – the self such as to subject it to "a persecuting possession", so that responsibility "could never mean altruistic will, instinct of 'natural benevolence', or 'love'"; "under accusation by everyone, the responsibility for everyone goes to the point of substitution. A subject is a hostage" (1991b: 111, 112). Confronted with the Other, there is in play "a debt preceding the loan", "an accusation preceding the fault", such that "the self is a *sub-jectum*; it is under the weight of the universe, responsible for everything" (1991b: 113, 116). Insisting that subjectivity is – comes into being, into effect, as – subjection and substitution, Levinas advocates a "modern antihumanism" which "clears the place for subjectivity positing itself in abnegation, in sacrifice, in a substitution which precedes the will. Its inspired intuition is to have abandoned the idea of person, goal and origin of itself. Strictly speaking, the other is the end; I am a hostage, a responsibility and a substitution supporting the world in the passivity of assignation, even in an accusing persecution, which is undeclinable". Humanism as we know it "has to be denounced only because it is not sufficiently human" (1991b: 127f.).

Reading these statements, it is no wonder that Paul Ricoeur speaks of *hyperbole*, meaning the "systematic practice of *excess*", in assessing the change of style as well as substance that strucks any reader comparing his first and second major ethical works (1992: 337). It is as if, for Levinas, the more "prior" to everything to do with the capacities for receptivity as facilitating agency in the self standing accused, being persecuted, hostage, etc., the better for responsibility as he came to conceive it; in the end, it is not even strong enough, hyperbolic enough, to assert that "a neighbor concerns me outside of any a priori", but "*before* every a priori, or from an earlier moment than that of the a priori". Indeed, he tells us, "this is the notion all our inquiry means to bring out, so as to reach the concept of an absolute passivity" (1991b: 192, note 20). Everything that goes to portray what happens to the self once the Other shows up is bestowed the mark of excess Ricour points to: separation between self and Other, exteriority and alterity of the Other, the passivity of the self – the word "absolute" is resolutely placed in front of all of them.

In an earlier work I wrote that "in the end Levinas so overemphasizes his case for the primacy of the Other that he underplays the ethical status of the acting I-subject to the point of the wildly implausible. The Levinasian ethics of the Other is attained at the price of the surrender of the I. Selflessness is nothing on which to build an ethical relationship with others" (Vetlesen and Jodalen 1997: 162), any kind of others, I now add.

To be sure, to make the point that there are various kinds of others for (towards) whom responsibility is imperative, and not only fellow human beings, is nothing short of an embarrassment at this stage. That said, the question here is to what extent this is an embarrassment, in the sense of a serious deficit, in the case of Levinas' ethics. But isn't this the very question I have already tried to answer? Wasn't my discussion of Derrida's reading of Levinas intended to settle whether there is an *animal other* in his ethics? Moreover, didn't I argue that there *is* a place for the animal when, in Chapter 2, I defended his essay on Bobby against criticisms from Deborah Bird Rose and Luce Irigaray?

I did, but since then we have gained a deeper understanding of Levinas' overall project. We have seen that his thinking about "the Other" underwent a radicalization to the point of hyperbole (Ricoeur), ending up in a series of claims about responsibility assuming the form of "substitution" and "being hostage", raising serious questions about where that leaves the self in terms of being at all able to respond responsibly.

In returning to the question of the place of animals, the worry is not about the dangers involved in emptying the self of the powers and capacities it needs to respond responsibly – a self-effacement bordering on what Freud called "moral masochism". There *are* grounds for worrying about this, but since we are now concerned with the self's relationship with non-human – meaning animal – others, the danger must be relocated, as it were; it pertains to *the wider world in which the self finds itself*, not simply to its partaking in a dyad.

To see where I am heading, the truly crucial statement by Levinas is the one cited above where he speaks about "the egoism of the conatus", about persevering in being as "egoism or Evil", outlining "the dimension of baseness itself" (1993a: 137). This connects with Levinas' notion of responsibility in that what evil is, or effects, is the "refusal of responsibilities" on the part of the self. Consider what is at stake. Levinas is locating the source of evil, amounting to the refusal of responsibility, at *the deepest possible level*, that of the very being of the self, its *conatus*, its striving to persevere in its being. That he is upping the ante in this fashion is testimony to his eagerness to make his case by way of radical excess. In other words, *simply to be and to have an interest in persevering in that being is the cardinal sin*, is what threatens the Other by way of undermining responsibility.

However, to think that this is just the familiar story – Christian or otherwise – rendered one more time, is to miss the point. What matters here is that such a view of what relationships are about and of what constitutes responsibility is profoundly anti-ecological, and for that reason also anti-ethical. Why?

To see why, recall the shift from the first to the second of Levinas' major works: from a situation where the self welcomes the other in an atmosphere of gentleness, nourished by an existence that is "good" thanks to the elements of nature (the sun, the water, the soil), to a situation where the self is said be the other's hostage, taking up space from the Other simply by existing, a space that Levinas gives us to understand is *meant for the Other*, not for the self. This is what is involved in locating the relationship between self and Other at the deepest possible level: the very being of the self imposes a threat to the Other, threatens to be at the expense of that Other.

Essentially, then, the relationship between self and Other painted in late Levinas takes the form of zero-sum, either-or: "absolute passivity" on the part of the self is imperative to ensure that it respects the "absolute primacy" of the Other. Were this self as much as an instant to be preoccupied with its own entitlement to being, to persevering in its own being, the Other's entitlement to being would be undone. The implication is that the world – or the relationship – isn't big enough for both parties, the self and its Other, regardless who that Other is.

Where does ecology come into this, not to forget the animal other?

The insight from which ecology starts out is that no creature can live without others. Life, going on being, is all about depending on others, not being self-sufficient in one's own existence, needing others to sustain it. In a word, life is about taking up space in a web of relationships with others to be respected as the *sine qua non* of staying alive. My "others" are not a threat to me, nor am I to them.

To heed this insight from ecology is to see that Levinas has it wrong about *conatus*: the one-to-one connection he postulates between having an interest in persevering in own's own being, termed "egoism", on the one hand, and evil on the other, is plain wrong. It is so not only ecologically, as a depiction of the conditions of existence, but also – more seriously for Levinas – ethically. It does not help to point out here that separating ontology and ethics is precisely what Levinas sets out to do, staking his entire case for the series of "prior-to's" that go with responsibility on that very separation, the more "absolute", the better.

The problem is not the obvious one that Levinas contradicts himself in that he does locate the ethical *Urszene* where the Other shows up for the self in a condition (or mode) of being-in-the-world, keen to show how drastically that mode of being is destabilized at that very instant. How could he not, somehow, locate it thusly? No, the problem I have in mind is even more internal to Levinas' stated aims. C. Fred Alford puts it like this: "What if the human [or ethical – AJV] problem were not egoism, but instead our terrible dependence on others? Not 'I want my place in the sun', but 'I need you in order to be me'" (2002: 22). The need for others Alford speaks about is existential as much as ethical, a condition of life that has to be met for any sort of responsibility for anybody else to be performed, making the inter-dependence of the self and the other(s), the going-both-ways of the relationships they form, a condition of ethics, not a threat to it.

Recall the first version of the self's being-in-the-world offered by Levinas. The impression created is that of abundance, richness, of having one's every need met thanks to the generosity in giving on the part of the elements of nature. For all its abundance, however, there is something conspicuously absent. No, I don't mean human others – that they are absent here only to show up later, like "intruders in the night", is an all-too-familiar point by now. It's the absence of animals I think about. Animals play no part in the abundance and richness of existence, of sheer being-alive, painted here; nor do they show up together with the (human) Other when he or she enters the scene. Whether it be early or late Levinas, animals make no difference, not existentially, not ethically. There is only Bobby, that lone dog, showing up in the camp during the war, only possibly – disputably, as we saw above – accepted as a member of the moral community, making it more-than-human, contingently not essentially so.

What difference, then, would taking animals into account make to Levinas' ethics?

In fact, the absence of humans' interacting with animals alert us to the absence of any sort of interaction-founded relationships *at all* in Levinas' ethics. The abundance and richness so striking in early Levinas is belied by a double absence, that of humans and animals alike. It is not a scenery of richness but of impoverishment. You might respond that the scenery is meant as a construct, much like Hobbes' "state of nature" or Rawls' placing his interlocutors behind "a veil of ignorance". True, but such constructs always rely on a certain degree of plausibility, otherwise they wouldn't succeed as starting points even for very general and abstract reasoning.

It is tempting to say the same thing about Levinas that I said about Habermas: that his omitting animals from what really matters, ethically speaking, mirrors and helps reinforce their decreasing presence as so many non-human others that play in role in making us human. Reading Paul Shepard, as we did in the first chapter, reminds us how historically recent is their lack of presence in the everyday lifeworld of most of us "moderns"; how recent is, also, the situation where cats and dogs make up the vast majority of animals we relate to when we do so. Shepard's question – "Why do men persist in destroying their habitat?" (1982: 1) – is just as little Levinas' question as it is Habermas'.

So what is lost when animals are kept out of the framework meant to address the ethical questions that matter, in a historical situation when, truth be told, their absence is more accurately described as taken for granted than as conspicuous?

What is lost, most fundamentally, is the connection between ecology and ethics: the insight that the conditions of life define the conditions of ethics, thus making it possible for life to go on for all sorts of living beings defines the task of ethics, its *raison d'etre*. The precise nature of this connection will be a major topic in the next two chapters.

To return to Levinas: where does the connection I talk about between ecology and ethics prove important in assessing his ethics?

Shepard reminds us that life is essentially about eating and being eaten. As Rose helps us recognize, this implies an ever-going exchange, alternation, taking-turns between living and dying. That somebody dies is the condition for others to go on living. Ecology is the study of the kinds of inter-dependencies between all sorts of creatures, and species, that this amounts to in real life.

Does this mean that the existence of the other is a threat to that of the self, any kind of self? In different ways, Hobbes, Sartre, and Levinas all suggest as much – though in Levinas the roles are reversed in that the self, *should it assert itself as a self* in the sense of claiming for itself the same right to live as the other, would threaten the other; therefore, it is ethically incumbent to renounce such self-assertion (more on this below). Either way, the essence of the relationship is that of conflict: one party occupies the place the other is entitled to. This is the either-or, the "ain't big enough for both of us" referred to above. It is an understanding of the ethical relationship that sits poorly with ecology.

V

Alford, in his important book *Levinas, the Frankfurt School and Psychoanalysis*, can help us see what is at issue here. Even though his analysis concentrates on relationships between humans only, his discussion sheds important light on humans' relationships to animals as well.

Alford points out that Levinas is afraid of relationships. Nearness – if at all – needs to be distant (as Habermas once said about Adorno's position). Why? Because the self's engaging in relationship with the Other risks breaking that other. Since the self's being at all concerned with itself, with its "perseverance in being" (*conatus*), is tantamount to "evil", response in the form of responsibility needs to efface the self's concern with itself entirely: for it to raise its ugly head would be to let it override responsibility, thus risking to destroy the Other. Separateness must be absolute and unconditional.

Appreciate that in saying this, the stakes are higher than in the case of the worry about reducing the Other to "the same", that is to say, the formula in Levinas that points to a certain similarity to Adorno's concern with the "non-identical". Of course, both thinkers are scared of totality, of totalizing tendencies that amount to symbolic violence against the particular. However, as the psychoanalytically informed Alford observes, Levinas' notion that the Other risks being destroyed by the self's engagement with it because failing to be uncontaminated by any dint of narcissism, of self-concern, is deeply misguided; it distorts the human condition and with that the conditions within which ethically charged relationships play out.

The purity demanded on the part of the self, to the point of hyperbolic excess, has as its corollary the frailty of the Other. Yet this frailty is exaggerated to a degree corrosive not conducive to the ethical task *par excellence*, responsibility. Why that is so is not easily recognized when reading Levinas, nor is it illuminated in the approach taken by Derrida. What is called for is bringing in a view that represents an alternative to Levinas' position, challenging assumptions central to it yet unquestioned so far.

The alternative view I have in mind is one that Alford makes constructive use of in his book, that of the psychoanalyst Donald Winnicott. To be sure, I have discussed Winnicott above, pointing out that his notion of "transitional objects" neglects the role of animals – Shepard's living animals at that, not stuffed ones. Turning now to his classic paper "On the Use of an Object", we find an understanding of the place of destruction, and more indirectly aggression, that makes for a highly instructive contrast with Levinas.

Winnicott's view is that "it is the destructive drive that creates the quality of externality", a view he means to counter the orthodox one that aggression is reactive to the encounter with the reality principle. He presents his view in the form of a sequence where

> The subject says to the object: "I destroyed you", and the object is there to receive the communication. From now on the subject says: "Hullo object!", "I destroyed you. I love you. You have value for me because of your survival in my destruction of you. While I am loving you I am all the time destroying you in (unconscious) fantasy." Here fantasy begins for the individual. The subject can now *use* the object that has survived. It is important to note that it is not only that the subject destroys the object because the object is placed outside of the area of omnipotent control. It is equally significant that…to say that it is the destruction of the object that places the object outside the area of the subject's omnipotent control. In these ways the object develops its own autonomy and life, and (if it survives) contributes in to the subject, according to its own properties. […] This quality of "always being destroyed" makes the reality of the surviving object felt as such, strengthens the feeling tone, and contributes to object-constancy. The object can now be used.
> (Winnicott 2018 [1968]: 222, 226)

How does this understanding of the constructive role of destruction in exploring (testing) whether the object (other) survives it, help us see features of Levinas' teaching that we haven't seen already? Alford puts it like this:

> Man's metabolism with nature, as Marx called it, it not confined to inhuman nature. We eat and breathe each other. What's needed is an ethics of limits and responsibility. Levinas is helpful here, but not as helpful as he might be because there really are no limits for Levinas. If I owe you everything, if I'm responsible for your sins, if there is nothing I dare withhold, then it actually becomes more difficult to think about limits, above all the limits that circumscribe the necessary ruthlessness of life. Infinity is not a limit. […] Levinas does not address this

> issue in the spirit considered here, *a morality of the inherent exploitation that is life.* If Winnicott is correct, life is inherently exploitative not just because we use others to survive, but because *in using others we discover that they are real.* Exploitation is the price of knowledge of the reality of others. Absent this knowledge, Levinas must work twice as hard to protect others, which is why he posits such a terrible distance between humans.
>
> (2002: 56; my italics)

Even though others are fragile, they are not, argues Alford in the spirit of Winnicott, quite so fragile as Levinas (and to a certain extent Adorno) believes. Often others are able to resist intrusion and destruction, and when they do, "joy happens: the subject feels in his or her bones the separate reality of the other, and so can engage and use the other in a responsible manner" (2002: 56); "use" understood, in Winnicott, as something creative and productive, not exploitative. As to the other's fragility or lack of such, then, pointing to limits, to responsibility, the ego *needs to find out*, to "test" the other as object in the sense of Winnicott's sequence. But this is precisely what Levinas forbids.

What Winnicott allows us to see is that Levinas is so marked by fear and trembling before the ego, at the thought of what the ego might do to the other were it to act from anything else than "absolute passivity" as brought on by being the other's hostage, that he (Levinas) never for as much as a moment will risk allowing the ego to "test" and thereupon use the other as an object in Winnicott's understanding – to risk that is tantamount to jeopardizing the very existence of the other. Hence everything to do with *the ego's need for the other* must, by all means, be suppressed, most basically its need to explore the realness of the other by way of attempting to destroy it (in Winnicott's non-literal sense of testing), such that what qualifies as real is what resists me, proving it exists separately from me. This is the difficult yet precious experience needed to make me realize that I am dependent on others who exist in their own right, pursuing their own projects, and so remain outside of my control, rendering my realization of being dependent on them psychologically demanding, calling for maturity.

So Winnicott helps us realize that every existence uses others, that is to say, uses them as objects in his intended sense, where such objects are real-life other human persons, in keeping with Winnicott's human-centred approach. As Alford rightly observes, the only ethically important question concerns "the limits of use, so that use does not become exploitation" (2002:105). He finds no satisfactory answer to this question in Levinas, searching in vain in his thinking for what Alford terms a sense of "shared freedom", a freedom that involves my participation with others, neither serving nor enslaving them, but living with them, acknowledging my dependence upon others, but not transforming this dependence into an idealized servitude, that is to say, what Freud called "moral masochism", assuming here the form of "glorification of guilt and subjugation for the purpose of erasing the self" (2002: 69). In psychological terms, the zero-sum, either-or I spoke about above, seeing them as the two extreme alternatives Levinas leaves the self with in its response to the other, resembles a borderline experience in its quality of "all or nothing at all", such that "either I am for myself, a narcissist imprisoned in my own world, or I abandon myself to the other utterly"; "either I impose myself and my frameworks on the other, what Levinas calls totality, or I open myself to infinity" (2002: 115). It is tempting to say – Alford does not, but he could have – that the stance Levinas paints with regard to how the self responds/relates to the other has more in common with what psychoanalyst Melanie Klein (1988, 1989) calls the schizoid-paranoid position than the depressive one, where the first disallows for the ambivalence and ambiguity tolerated in the second, tolerated in the self (myself) as well as in the other, acknowledging that *both* self and other are capable of *both* good and evil. To tolerate these qualities in oneself as well as the other is important to counter the fear that the other would shatter should I come too close, should I try to come close at all, close enough to *get it wrong*, to contaminate the other with my badness, my greed, and my envy; close enough to hurt and to do harm, or to neglect and forget, creating the need for, and trust in the possibility of, reparation, implying forgiveness, and gratitude at the willingness to forgive. These key aspects that make up the human repertoire in relating/respond to the other are largely absent in Levinas. It is the price he pays for exaggerating the presumed frailty of the other.

What about love? Alford quotes Levinas' statement "Eros, strong as death", taking him to mean that "love, like death, seeks infinity. Because the other is so other, to love another is like death" (2002: 70). This goes deeper than, yet reminds us of the French saying that every time you go, I die a little. However, I disagree with Alford that this is the most important feature in Levinas' thinking about death.

Common sense has it that death is the ultimate threat to love. What I am after here is something else, though. It is to do with the implications of bringing the psychoanalytic insights of Winnicott to bear on the constellation ethics-ecology in Levinas.

Recall Levinas' statement that "the being that perseveres in being, egoism or Evil, outlines the dimension of baseness itself"; that "the ego, responsible despite itself, abrogates the egoism of the *conatus*" (1993a: 137, 138). That conatus, an entity's striving to persevere in its being, should be identified (one-to-one) with egoism is highly contestable. To occupy a place in the world, to persevere in being the kind of being an entity, be it the human person, is, is not something that threatens the possibility of responding/related to the other, to be and do something that matters to that other, but its precondition. If the reply is that Levinas is out precisely to locate the self's response-responsibility *prior* to being – to being for-itself as well as for- or with-the-other – my rejoinder is that what is problematic is precisely this situating response-responsibility *before* being – before not simply in terms of temporality (chronology), in itself a big topic in Levinas, but before in terms of primacy, making it imperative for the self to abandon its claim to being, its conatus ("abrogate" its egoism), so as to give all the space to the other, claiming none for itself.

The upshot of what I am saying is that the "egoism" of the conatus is not tantamount to evil but the precondition for making the very kind of choice that defines ethics, that between good and evil: rather than evil being implicated in the conatus as such, it makes the choice of goodness – the alternative to evil – possible.

Then there is destruction, in Winnicott's non-literal sense, explained above. Because of his fear for the intrinsic frailty of the Other, a fear whose corollary, we now see, is his fear of the interest-in-persevering in the self (ego), Levinas is scared to death when it comes to destructiveness – scared, that is, that were the urge to destroy the other in Winnicott's sense of testing what the Other can take to play out, if only for an instant, death would follow, if not literally, than ethically: the chance that the Other would survive any such use by the self can never be taken, is never worth taking.

To be sure, in advancing this critique I violate the proclaimed first principle in Levinas' whole approach to ethics: that ethics is prior to psychology. To grant that premise and the extreme purity it involves, however, comes with the price of emptying the self of all resources needed to relate to others in a responsible manner. To refer like I have done, then, to the self as responding/relating to the Other, is already to disobey Levinas' premise that response in the form of responsibility be prior to, and thus clearly separated from, any form of relating with that Other, suggesting as relating does the participation, of both, in a relationship. Everything to do with interaction, being-with the Other, comes *after* the self's being-for the Other, after both chronologically and ethically, according to Levinas. That is why the lack of a sense of "shared freedom" that Alford charges Levinas with, a freedom that involves my participation with others, is no deficit for Levinas but key to his project. This explains Levinas' criticism of Martin Buber's *I and Thou* – namely that Buber does not fully break with the traditional privilege of ontology; his failure to do so being evident in his characterization of the I-Thou relation in terms of being, instantiating a "between" (inter) that is "a mode of being: co-presence, co-esse" (1993b: 23). Hence for Levinas Buber's thesis that "in the beginning was the Relation" shows that he has got it wrong from the start. To conceive of the I-Thou as a relationship in the form of a dialogue is to situate it on a footing of equality, of reciprocity, and thereby to jeopardize what is distinctive to response-responsibility in Levinas: the I's giving without expecting anything in return.

Let me conclude in a way that anticipates things to come. There are three points to make, one affirmative, the others critical.

First, and foremost, the truly crucial takeaway from Levinas' ethics for my purposes is its insistence that the ethical relationship is asymmetric, that responsibility is not conditional on reciprocity, and that the Other's vulnerability is what grounds my response qua responsibility. The susceptibility to suffering that this vulnerability implies is such as to include animals and other sentient beings as my addressees. Indeed, it is tempting to say that the asymmetry Levinas insists on suits the human-animal constellation *better* than it suits the human-to-human one (a point I shall elaborate in the next chapter). It is not necessary to be an agent capable of responsibility in order to qualify as an addressee; vulnerability suffices.

Second, encounter has primacy over knowledge, yet every encounter between a self and an Other, human or animal, involves both parties in a relationship with each other, within which alone responsibility can be enacted in ways that may prove right or wrong, good or evil. This being so, Levinas' insistence that responsibility's being asymmetric depends on its *separation from being*, from the dynamics of relationships, is misguided and comes with unacceptable costs, situating the responsive self in a psychic and hermeneutic no-man's land, casting psychology as detrimental not co-constitutive for the attainment of maturity that is needed for any context-specific enactment of responsibility.

Third, there is in late Levinas a penchant for hyperbolic excess that results in the self's effacing itself so as to do justice to the Other as face. This is ethically counterintuitive and psychologically counterproductive. It involves a self-imposed suppression of engaging with the Other creatively in Winnicott's sense of destructiveness, thereby imprisoning the self in a mechanic (rigid) as distinct from organic (flexible) stance where giving space to the Other comes with the masochistic cost of claiming none for itself, not even the space needed to do what responsibility for the other demands, namely a minimum of resources on the part of the self, lest the appeal from the other go unanswered. And by misconceiving conatus as egoism and primitive narcissism, as always *at the expense of others* instead of as facilitating the self's playing a vital part in their sustaining themselves, Levinas misses the chance to fuse his ethical vision with ecology, with the wider world. The present-day rhetoric promoting that the human agent minimize or give up his or her voice for the sake of so many voiceless other-than-human others, ashamed at being (and having for so long in Western history been) the centre, may well be the legacy of Levinas' ethics that regrettably proves most influential within contemporary posthumanism, taking his proclaimed antihumanism in a direction he did not, and perhaps could not, foresee.

4 Encounters as experiences of conversion

I

On February 19, 1985, the Australian ecophilosopher Val Plumwood experienced a life-changing encounter in Kakadu National Park. Alone in a canoe and about to return to the shore, she noticed what she took to be a log floating in midstream. It turned out to be a saltwater crocodile. Her attempts to frighten it off did not prevent the crocodile from tightening its jaws around her legs, dragging her down under the water in what she came to describe as "three death rolls". Despite severe leg injuries, she managed to break free and reach the shore, then walked and eventually crawled the three kilometres to the range station, where her cries for help were heard and she was taken to hospital, staying in intensive care for a month.

Forty-six years old at the time of the incident, Plumwood was already a widely recognized figure within the emerging field of environmental philosophy, a status that would be galvanized by the success of her first book, the path-breaking *Feminism and the Mastery of Nature*, published in 1993. Surprisingly, though, the experience with the crocodile is not a topic in the book, not explicitly. Only about ten years after it occurred did she start writing about it, having taken her a long time to digest and be able to reflect upon it in a philosophical manner.

The following is an excerpt from Plumwood's article "Taken by a crocodile", published in *The Age* on January 12, 2004:

> Before the crocodile, I wrote about the value of nature, but after the crocodile, I started writing about how we see ourselves as outside nature, about the power of nature and our illusions that we can control it, that we're not embodied beings and are apart from other animals. We don't understand ourselves as ecological beings that are part of the food chain, where we must honor our food and the more-than-food-that all of us are, including other life forms. I don't believe we do this when we treat other animals as no more than food.
>
> (Plumwood 2004)

DOI: 10.4324/9781003317692-5

In "The Eye of the Crocodile", she addresses the incident's profound impact on her thinking about humans and animals:

> In the individual justice universe you own the energy volume of your body absolutely and spend much time defending it frantically against all comers. Any attempt by others at sharing is regarded as an outrage, an injustice that must be resisted to the hilt. In the other, Heraclitean universe being in your body is more like having a volume out from the library, a volume subject to more or less instant recall by other borrowers – who rewrite the whole story when they get it. [...] The crocodile event changed my emphasis to one of better understanding our dependency on nature and our over-estimation of our powers. We need to understand nature because we have a sense of security that is illusory. The dominance of rationalism in our culture creates a degree of over-confidence.
>
> (2012: 35)

What Plumwood describes is a fundamental shift in her thinking, a shift provoked by how the crocodile regarded her, namely as food. Her instant reaction, she says, was that the crocodile had it wrong, utterly wrong to think that she could be reduced to food; as a human being, she was so much more than food. The experience with the crocodile forced her to acknowledge the existence of a world "from the outside", a world no longer her own, no longer the world as defined by, and for, humans but instead "an unrecognizably bleak landscape composed of raw necessity", a world that would go on without her, indifferent to her will and struggle, to her life and death (1996). The "hyper-separation" of humans from nature that she so powerfully critiqued in *Feminism and the Mastery of Nature* was given the lie in most dramatic, autobiographical fashion ripe with philosophical implications, imposed upon her by an animal predator who subjected her to an experience only non-human animals, not humans, are supposed to have, namely that of being hunted, not the hunter, exposing her to the terror of being the Other's food, nothing else.

Plumwood calls the world from the outside that the crocodile forced her to acknowledge as *its* world, as real as hers, undoubtedly, indisputably so, the Heraclitean universe. Being confronted with its realness the way she was made her realize how it happens that one predator recognizes another predator; how the roles of hunter and hunted alternate between the species and so may suddenly be reversed between them, according to the dynamism of a universe from which humans are not – precisely not – exempted. To think that we are, on the basis of some capacity-related "superiority" vis-à-vis the rest of nature that philosophers as well as Western society at large have insisted upon, is *hubris*. It attests to an over-confidence in us humans that is not simply wrong, meaning not true to reality, but also extremely dangerous, not least for us and our prospects of survival qua species, nothing less. What this over-confidence gets wrong is wrong not only, or even primarily, for ethical reasons. At the fundamental level to do with the factual conditions of life, what is wrong is the notion that humans are independent of the rest of nature, and that nature is there for the taking, passively awaiting whatever we, in our freedom and autonomy, may choose to do with it or, more to the point, against it. As a culture, we hold ourselves to be "free" to act *as if* non-human nature in its richness and entirety exists as a means, as so many resources, for our purposes, inducing and justifying us to treat as substitutable what *is* indispensable, as unlimited what *is* limited, and so on. Since this *as if* is in actual fact not true to the conditions of life of which human life is a part, unconditionally and ineluctably at that, pursuing the as if of superiority and independency on our part is not going to "overcome" or in any way neutralize the realness of these factors – quite the opposite, it will only amplify and exacerbate their impact, making us more vulnerable for each year, each day we go on denying the sources of that vulnerability.

The denial of our dependence on nature is a big topic in its own right, one that I have addressed in *The Denial of Nature* and that Plumwood examines in what was to be her last monograph, *Environmental Culture: The Ecological Crisis of Reason* (2002). Indeed, the title gives away her point: What has passed as "reason" in hegemonic Western thought, to the point of pervading the major structures and institutions of modern society to this day, is on closer scrutiny exposed as a series of blind spots that, precisely in being *acted upon*, now increasingly reveals the reason that produces them as "actually the opposite of reason" (2002: 118).

For my purposes, the most relevant aspect of Plumwood's analysis of how denial of humans' dependence on nature plays out is to do with our relationship with animals, including crocodiles. The one particular blind spot deserving mention here is that involved in promoting "insensitivity to the Other's needs, agency and prior claims as well as belief in the coloniser's apartness, superiority and right to conquer or master the Other" (2002: 118), what I have elsewhere referred to as a sense of entitlement, the right to have it all (Vetlesen 2021). The promotion of insensitivity, Plumwood writes, is the very function of the human-centred mentality, "centrism" for short, that she talks about. Insensitivity towards non-human others' needs and agency is dangerous in that it "puts the centric perceiver out of touch with reality about the Other" (ibid.). Such insensitivity must be fought on all fronts, in all areas where it currently prevails, the positive objective being to "establish better communicative relationships with nature in all its aspects, as a preliminary to learning to balance human needs with nature's needs and limits" (2002: 142).

This is Plumwood's general outlook. But where does she stand with regard to animals? Does she favour including them within a moral universe of rights? Are animals carriers of intrinsic value? Are they "ends in themselves" (the status Kant reserved for humans) that would be violated were we to treat them as food and to eat them?

Without engaging extensively with the supporters of animal rights inspired by Peter Singer's work in particular, Plumwood makes it clear that she finds the extension of the rights framework to the non-human world problematic. Why? Because the concept of rights has too many affinities to the human-centrist framework she sets out to challenge and, hopefully, help replace with an alternative more true to the ecological facts just mentioned. For one thing, it is difficult to apply the concept of rights to non-humans "beyond the case of certain animals that resemble humans"; indeed, the notion of rights is committed to the "neo-Cartesian moral dualism", she critiques, a dualism that organizes moral concepts and the status of the entities they are meant to pronounce on in hyper-separated, all-or-nothing ways, such that a being *either* is ascribed a full-blown right to treatment considered equal to human treatment, *or* it is not considered entitled to any form of ethical consideration at all (2002: 144). The categorical manner in which, say, Kantians make the distinction and draw the line (sets up a barrier, a wall) between those inside and those outside the universe of what I have called "moral addressees", so that *either* an entity (human, animal) is treated as an end in itself and hence inviolable, *or* as a mere means, is *part of the problem*, leading to what in Plumwood's terminology is the hyper-separation between "use" and "respect", conceiving the two categories as mutually exclusive, something we shall see is as deeply problematic in practice as it is in theory.

The concept of rights is also problematic for philosophical reasons of the kind I touched upon in my discussion of Habermas in the previous chapter. We saw that the moral universalism which Kant-inspired contemporary thinkers like Habermas adhere to is wedded to a methodological, ontological, and moral individualism ill-suited to address the status of non-human beings, including the factors that undermine it, revealing a near complete lack of being in sync with current ecological realities. The upshot is that relationships considered with regard to their moral import are indiscriminately forced into the one-size-fits-all homogenizing straitjacket of the one-to-one constellation, such as to answer, in simple terms, the question of who is the agent and who is the addressee. Thus conceived, everything relevant to determining *who is what* in the particular case or situation is, first, sought answered in terms of capacities, whereby, second, everything seen as relevant to decide about responsibilities, obligations, and rights are taken to be qualities owned by the individual entities in question *qua* individual agents and/or addressees. As I pointed out in my critique of Habermas, this individual-centredness comes with the price (among others) of silencing the extra-individual, structural and macro-related forces and constraints that impact on the decision-making powers – in a word: the actual agency – of the individuals involved. The result is that environmental problems such as climate change per conceptual fiat have to be reduced to the decisions and actions of so many discrete individuals in some discrete spatio-temporal settings in order to be made morally assessable and legally accountable, so that the whole, complex problem (anthropogenic climate change) is regarded as an aggregate of the actions of so many individuals, neglecting the impact of structures and institutions much like neoclassical economists are wont to "externalize" them, and appealing to the moral character of individuals in order to "solve" the problem.

Plumwood's critique of the advocacy of animal rights is similar to my critique of Habermas. She traces the individualistic tendencies of rights theory to the way it "takes a complex set of social and political conditions and congeals them into a singular and apparently simple and stereotypical quality of an individual whose rights are violated by another individual who is responsible for their violation" (2002: 153). To be sure, these are general statements about conceptual matters and general positions. What we need to do is bring them to bear on the particular issue at hand, brought so dramatically to the fore in Plumwood's eminently one-to-one encounter with the crocodile: the issue of how animals should be treated by humans, with special attention to their – and our – status as beings that can be food for their species "others".

Having pointed out the overemphasis on personal conviction that she sees follow from the individualistic premises of rights theory, Plumwood contends that this theory "leads to the adoption of poorly contextualized and culturally hegemonic vegetarian strategies which condemn all eating of animals in indiscriminate terms, and do not encourage us to prioritise opposing the most extreme examples of distortion and instrumentalization of animal lives – the intensive farming practices that treat animals as no more than living meat or egg production units" (2002: 154). These are practices that effectively reduce, say, feedlot salmon to *biomass*, denied any agency and individuality within a regime where humans decide everything to do with the total life cycle of the fish, that is to say, in the description given by the philosopher Martin Lee Mueller, "the circumstances under which the fish will be fertilized and grow, what they will be fed, where and how they will die, and who receives their flesh afterwards" (2017: 39). In other words, the poverty of rights theory is not limited to the way in which it individualizes the parties whose moral considerability and responsibility are assessed, rendering morally motivated resistance and activism largely ineffectual for lack of engaging head-on with "the responsibility of systems of economic rationality for the atrocities daily committed against domestic animals, especially in the factory-farming framework" (ibid.), demonstrating that this sort of individualism is misguided also in terms of carrying real political traction.

The analytic poverty Plumwood alludes to goes deeper than this, though. Her point may be introduced by reference to my experience when presenting Deborah Bird Rose's book *Wild Dog Dreaming* in my class of master students. As recalled from Chapter 2, hunting is a central theme in the book, especially – literally - hands-on in the section where Rose writes: "Often an animal would look at us in the moment before it was shot. We saw it die; sometimes we had to finish it off with our bare hands. [...] When you cut into an animal, the thing that is so surprising is the reminder of how warm animals are on the inside" (2011 25f.). Upon my reading this passage to the class, there was a long silence in the room, tensely and a bit awkwardly so. Hesitant at first, the comments that eventually ensued were testimony to the surprise, even shock that most of the students clearly felt: they had not expected the author of a book in their environmental philosophy course to endorse the killing of animals that takes place in hunting, let alone to participate in it: Perhaps Rose's being a woman amplified the shock.

It will not come as a surprise to the readers of this book that being against humans' killing of animals is the taken-for-granted position among, say, the present generation of Norwegian philosophy students, as well as among what seems to be an increasing portion – albeit still a minority – of society at large. Especially within academia, the burden of argument is placed one-sidedly on the part of those who, for some reason or another, are in favour of the killing of animals for human purposes – reasons that are assumed either to be tied to economic self-interest or to a dubious moral and/or political outlook, or a combination of both. Hence the opposition to killing animals has an air of moral superiority, representing the enlightened and civilized stance.

I take the class room experience to illustrate the point on Plumwood's part that I am getting at: one of the ways in which the animal rights theory is poor is its penchant for settling the question of the killing of animals in terms of an either/or, only to advocate the position that the killing of animals for human purposes is *wrong*, period. The case is effectively closed, the issue resolved, once and for all, *sub specie aeternitatis*, in a manner priding itself on being exactly that: principled and universal, allowing for no exceptions, loyal to the Kantian exemption taboo such that no case or situation will allow for a different view or course of action, be it on pain of the worst sin of them all, that of self-contradiction.

Again, this is my way of putting the point, not Plumwood's. But it is a point we share. The question that must be asked is this: Is the question of whether it is right or wrong for humans to kill animals one that allows for, let alone requires, an either/or sort of answer, such that either it is right, meaning always and in principle, or it is wrong, always and in principle? Why should this absolutist approach, feeding on binary and allegedly mutually exclusive answers, this strict rigour in thought and inflexibility in action, be the approach best suited to the matter? Is the phenomenon as such – humans killing animals – really that homogeneous, that void of complexity, ambiguity, of reasoning of the "it depends" sort, of the experiences made by others, besides oneself, meaning not only other individuals but also other cultures, other cosmologies? Can the question be stylized into a purely principal matter, doing without the input from past history, from current culture, from the particularities of context? Into a matter that, in keeping with this logic, is actualized in one particular constellation as far as the parties are concerned, namely that of the individual agent on the one hand and his or her animal counterpart (addressee) on the other?

Given these criticisms, what would an alternative position look like, one allowing for greater flexibility and for the exercise of experience-based judgement on the part of the human agent, yet without forgetting to take into account the impact of extra-individual structures? A position that would welcome the factual heterogeneity of all the parties who, in some way or another, have their lives impacted by the approach taken and the options it either facilitates or excludes?

Eager to offer such an alternative position, Plumwood is adamant that it "would not place the emphasis on personal purity, and it would resist the idea that predation is inevitably fallen"; moreover, it would "refuse to create an absolutist moral boundary between animals and plants, as a dualism of conscious and non-conscious life in which only the former deserves moral consideration" (2002: 155). In keeping with my discussion above, the position sought for must reject the exclusionary imperative that derives from "the reliance of rights-based forms of vegetarianism on the (unstated) assumption that only those beings not admitted to the class of right-holders can ever ethically become food". Indeed, to equate availability as food with moral exclusion is to be (however tacitly) committed to the moral dualism that is a legacy of Cartesianism; it is a view that is forced to insist on "a substantial outclass of living beings that are morally excluded in order to locate any viable form of eating which allows an ethical basis for human survival" (ibid.).

The deeper problem with the exclusionary imperative stems from its commitment to conceptually erecting, and thereupon policing, boundaries between humans and animals (or some animals, not all), or between animals and plants, boundaries that are artificial, meaning constructs, and as such *imposed on* the domain of planetary life rather than *originating in* that life, reflecting it and matching its dynamics. Plumwood writes about the "intensification of the use/respect boundary around dualised forms of moral privileging that deny the ethical continuity of planetary life" (2002: 155), which is fine. But the point goes deeper still, since what is being denied by the endless exercises of boundary drawing on the part of just one species, and "on behalf of" all the rest, is not primarily the ethical continuity of planetary life, but its *ecological continuity*, a continuity in the strongest possible sense, namely that of cross-species interdependencies, where some are the food for others, the death of some the precondition for the life of others, and where the very distinction-cum-boundary between the living and the non-living is far from clear-cut, far from allowing for the inclusion/exclusion binary that is so often invoked (in moral philosophy, not in ecology!) as were it razor sharp. In the previous chapter, I showed how both Habermas' and Levinas' ethical positions, albeit for very different reasons, suffer from a blindness to the way in which *all* relationships and responsibilities – not only those among humans – depend on the intactness of a more-than-human natural world and the multitude of non-human species and life forms it consists of – a reminder of the earthly, including bodily and animalistic, dimension of humankind's existence ignored only at our peril.

But where does this leave us in terms of moral orientation? Accepting that we have to do without sharp boundaries between different species, and that the exclusionary imperative has to be replaced by a more flexible, context-sensitive, and ecologically attuned approach – attuned to interdependencies and shared vulnerabilities, that is – it cannot do without the drawing of distinctions altogether, can it?

Of course, it cannot. That being said, I shall not be able to address here the issue of appropriate, ecologically informed distinctions all across the board, but shall concentrate on those that are crucial to the main category under discussion, that of animals.

That animals form as richly heterogeneous category is an important fact, one too often neglected by philosophers, but at this juncture it does not take us beyond positing a point of departure. We recall that Plumwood faults animal rights theory for condemning all eating of animals in indiscriminate terms, giving "hegemonic vegetarian strategies" as her example (2002: 154). The "extensionist strategy" (think of Singer's "expanding circle" of moral considerability as a model), she argues, is committed to what she labels the Use Exclusion Assumption that "because food is inevitably a site of domination, degradation, and exclusion, ethical food practice must consist of ensuring that nothing that is morally considerable can ever become our food or be ontologized as edible" (2002: 156). Although this may appear to be a perfectly plausible assumption, such a restriction of what may be considered as edible or useful – useful in the sense of being used in a morally acceptable way – "results in a deep rejection of ecological embodiment for those beings, since all ecologically-embodied beings are food for some other beings" (ibid.). This is a vital observation, and it sits well with the point I just made about how what is morally appropriate must fit with the ecological realities of the beings (parties) involved. It also resonates with the insights I highlighted in the work of Shepard and Rose in earlier chapters.

Surprisingly, there is no mention of her encounter with the crocodile in Plumwood's *Environmental Culture*, published almost 20 years after it happened. Consider the following passage:

> The concept of meat justifies oppression by hiding responsibility for death and the causal connection between the production of meat and the animal's death. The backgrounding of erasure of these connections in the abstractly quantitative and commodified concept of meat [Carol] Adams terms "absent referent". "Absent referent" renders unavailable not only the act of killing which makes meat available as a commodity, but any recognition of the connection between the meat and the communicative being that it once was. [...] Such a commodified concept of

> meat involves a strong, fetishistic form of instrumental reductionism, in which the other is defined in narrow ways that identify them with what is only a part of their being, the part that is of use to us as flesh, and do not recognize that all living beings are much more than that.
>
> (2002: 156f.)

Why cite this passage in the light of Plumwood's encounter with the crocodile? Recall her reflection on the experience: "It was utterly wrong to think that I could be reduced to food. As a human being I was so much more than food". We may say that what happened when she was attacked by the crocodile is that she, a human being, was subject to being treated the way humans typically treat animals, namely to be identified with what is only part of her being, the part now demonstrably seized upon and used as flesh by the other, in this case a crocodile, not recognizing that all living beings are so much more than just flesh (food) for others. What happened, then, is that the crocodile turned the tables on the hegemonic understanding of the relation between humans and animals as far as "being food for some other beings" is concerned: it acted in a way effectively reversing the roles, something Plumwood the radical ecophilosopher was forced to realize she was simply unprepared for.

To actually make the experience of being reduced to food for an animal is no doubt as rare as it is shocking for a modern human being: few make that experience, and even fewer survive to reflect on it ever after. In premodern times, as well as in some indigenous cultures existing today, the experience is not only less rare; it is also integrated into the cosmology and mentality of the people, exemplarily by the participation in hunting, as we saw in Chapter 2 when discussing Old Tim's relationship with the dingoes as depicted by Rose. Presently, however, the large majority of us have been almost completely alienated from practices where humans and animals encounter each other in the roles of predator and prey, whereby the human never knows for sure whether he or she will end up as the prey, and the animal as the predator that prevails. The context-dependent attunement that hunting relies on, together with the respect felt for the animals hunted, acknowledging their significance as food needed for survival, attests to a way of relating to animals that is largely lost to us.

II

We need to probe deeper still, though. Why precisely is it "utterly wrong" to think that a human being could be reduced to food for some other beings? What happened was that the crocodile acted *as if* Plumwood was nothing but food, nothing but a means to its ends. If the crocodile is "utterly wrong" in treating her like this, what exactly makes it wrong?

I take it that nowadays the most common, indeed obvious answer to this question is not available to Plumwood, for so many reasons, be they philosophical, ethical, and more – namely that what is wrong is simply that the animal here assumes the role exclusively restricted to humans: it treats a human being not as an end in itself but as a mere means for its own ends; and this is something only humans can do, not to other humans, mind you, but to animals. So whereas in the former case, something is wrong, in the latter nothing is wrong, rather everything is as it should be, in keeping with current intensive farming practices. To accept that this is the correct way to view the distinction between morally right and wrong thus takes the form of adhering to the boundaries between species, understood in the categorical manner Plumwood dismisses.

Since the answer representative of hegemonic anthropocentrism is not available to Plumwood, what kind of answer is?

I believe that part of the answer is contained in the following statement: "There is injustice for any striving and intentional being in being constructed in reductive terms as an ethical nullity"; "instrumental conception defines the Other in terms that assume the right of a 'higher' group to treat them as a resource for their ends" (2002: 157). The context is that of animals used in intensive farming, very different from that of Plumwood's encounter with the crocodile. But the point she makes certainly speaks to her own experience as well, the key phrase being "the injustice for any striving and intentional being" in being reduced to "an ethical nullity". In the case taken for granted as the standard one, the being she refers to would be an animal, one violated in being reduced to such a nullity by human agents. Only in her own case, *she* is the striving and intentional being that is nullified in being reduced to food for another, and nothing more.

Is what is wrong, then, that the crocodile, assuming the role of the agent who calls the shots, as it were, violates her status as an end in itself, as someone who is not a mere means for its purposes? To answer in the affirmative would be to give the crocodile the compliment of being a full-blown moral agent, one capable doing right and wrong, and so blameworthy when acting wrongly. I don't think that actually captures what Plumwood would answer. I assume that she would be prepared to grant the crocodile many of the powers of agency that our culture routinely denies animals and uphold as uniquely human such; if anything, the way it attacked her was a demonstration that the crocodile's powers of agency in crucial (life-and-death-deciding) ways outmatch her own: she was clearly the inferior party, the overwhelmed and outmanoeuvred one. So agency, yes, but moral agency, no, because the capacities required to make a deliberate choice between the *morally* right and the *morally* wrong action to take, are of a kind only found in humans. Indeed, this is a key reason why humans have responsibilities towards animals, responsibilities not requited on the part of the animals towards humans, this being (among other things) what makes the relationship between them asymmetric. The point is not only that human agents are to be considered as responsible agents in a way animals cannot. It is also that one of the features distinctive about human life is to do with how we as persons have responsibility for cultivating, for not ruining or allowing to lie dormant, the capacities required for exercising such responsibility for (different kinds of) others. As Alice Crary observes, "individual human beings can bring themselves to states of decay with a moral ugliness that has no exact analogue in the lives of [animals such as] horses"; we have an ability qua humans to "injure ourselves by stunting our own emotional and rational development" (2015: 223, 232), decisive as such development is for our trustworthiness and maturity as agents in matters moral. To stress the individual's responsibility for cultivating the capacities required to enacting that responsibility is not to deny that various extra-individual – cultural, social, ideological, material – factors also play an important part.

Having established that what she found "utterly wrong" is not, given my argument, that the crocodile treated her in a morally blameworthy way, what does the "wrong" consist in? It must be linked to her being terrified by how this animal treated her; terrified in an existential way rather than an ethically oriented one, if the distinction makes sense here. She was afraid that she would die, that she would succumb to the physical – but not only physical – superiority of the animal, the way it was calling the shots, and *humiliating her sense of being humankind subjected to this by an animal*, an animal that for its part may well have thoroughly enjoyed doing so, thriving in the brilliant display of superiority that must have been equally clear to both of them. Was this animal playing cat and mouse with her? Turning her from sovereign human to a plaything for its enjoyment as long as it wanted to? Removing her from the pedestal of human supremacy she philosophically speaking is highly critical of, yet cannot help being shaped by, perhaps to a greater degree than she is prepared to admit? And when she eventually emerged alive from the encounter, she did so not because she outsmarted the crocodile, but because *it let her go – chose* to let her go, in a way demonstrating its superiority precisely by that choice, adding insult to injury by signalling that "you should be grateful that I let you go, it being completely up to me to decide the outcome of the encounter between us". *That*, I suggest, is the true depth of the experience Plumwood made, it being no less deep for not being essentially about moral agency, but about other dimensions of the agency of that animal, dimensions attesting to the ways in which the animal can be the active party, the human the one who tries to respond, as best as it (she) can and all the while at the mercy of the animal.

Notice how I put this point: In saying "it let her go", only the human party to the encounter is granted a sex, the animal is not: the crocodile is referred to as "it", not as "he" or she". Isn't this species-specific linguistic practice – per convention – a sort of discrimination that speaks volumes about how the cultural legacy of anthropocentrism still pervades our manner of thinking about animals? Rarely is the mismatch between "word and object" (Quine) as blatantly demonstrated as in this case: Plumwood's experience is all about the powers of agency of her animal counterpart, the crocodile. Yet linguistic convention *denies* rather than affirms that agency in referring to the crocodile as an "it", and so as belonging to the same category as nonliving, inorganic things like a table or a stone. This poverty of language, and language's role in perpetuating the ontological and epistemic poverty in question, is staggering. Not only is the animal's agency denied; so is also the possibility of recognizing the relation between Plumwood and the crocodile as an "I-Thou" as distinct from an "I-It" relation, to use Buber's distinction.

To return to the encounter: this is as far as I am able to go with what Plumwood does not say, but might have said, concerning how she came to understand what she experienced. For her, it was a "life-changing" experience, dividing her life into a "before" and "after" the crocodile. Thus it invites comparison with what the author John M. Coetzee calls "a conversion experience", much discussed among philosophers such as Stanley Cavell, Cora Diamond, Cary Wolfe, and Ian Hacking, following the publication of Coetzee's novels *Disgrace*, *Elizabeth Costello*, and *The Lives of Animals*. How does the notion of "a conversion experience" associated with Coetzee's work compare with Plumwood's experience?

III

Coetzee's book *Boyhood*, classified as nonfiction but no doubt in large part autobiographical, contains a scene where the main protagonist, a young boy, witnesses her mother's treatment of three hens supposed to provide eggs for the family. Failing to lay the expected eggs, this is what happens to the hens:

> His mother consults her sister in Stellenbosch, who says they will return to laying only after their thorny shells under their tongues have been cut out. So one after another his mother takes the hens between her legs, presses on their jowls till they open their beaks, and with the point of a paring-knife picks at their tongues, The hens shriek and struggle, their eyes bulging. He shudders and turns away. He thinks of his mother slapping stewing-steak down on the kitchen counter and cutting it into cubes; he thinks of her bloody fingers.
>
> (Coetzee 1997: 1f.)

Quoting this passage, Ian Hacking comments: The boy "sees what he sees, right there, but *he also sees what he did not see before:* the steak, bloody; his mother's fingers, bloody; the flesh and blood of something that was once alive, like the three chickens in the yard. He does not see it 'as' blood. He sees it for what it is, blood. The experience stuck with Coetzee all his life. There is no new information, but his eyes are opened" (2008: 147). This leads Hacking to suggest that, instead of speaking of the "difficulty of reality" (a notion made famous by Cora Diamond), we should speak of "the difficulty of experienced reality, of reality as experienced" (2008: 153). In a discussion with philosophers, Coetzee himself states that in regard to how we, collectively, treat non-human animals, in each of the participants "there took place something like a conversion experience", one that "as often as not centered on some ... mute appeal of the kind that Levinas calls the look, in which the existential autonomy of the Other became irrefutable – irrefutable by any means, including rational argument" (Cavalieri 2009: 89). He goes on to talk about friendship, pointing out that "friendship is rarely governed by reason. One does not choose one's friends according to rules. Becoming friends is like falling in love: one follows one's heart". And then he turns to our relationships to animals: "We cannot treat animals such as sheep and cattle as we would treat old clothes, discarding them when it is necessary (e.g., when we have all become vegetarians). We have, in effect, brought them into our world; we cannot conscientiously expect them to leave it, for there is nowhere else for them to go" (2009: 90f.).

I am not concerned here with the philosophical merits of Coetzee's case for vegetarianism, expressed most forcefully (many will probably agree) by his protagonist Elizabeth Costello. Rather what interests me is the nature of what Coetzee calls a "conversion experience". What kind of experience is it?

For a start, it is an experience in the emphatic sense of changing how the subject sees the world, and as part of that understands his or her role in that world, such that "after that happened", I no longer... Things come to matter in a different way than they used to, whereby their mattering differently is not initiated by the subject but instead something he or she undergoes, is affected by – thus the affinity between experience, negativity, and suffering that, in philosophy, goes back to Hegel and his take on the Greek tragedies (see Hegel 1977 [1807]; Gadamer 1975 [1960]; Alford 1992). Why suffering? Because the change in my world that is not of my own making may be painful in the way if offends a certain sense of autonomy, of determining what matters, and how, rather than being at the receiving end of the mattering, being literally hit by it so as to never be the same person, inhabiting the same world, afterwards – a different self in a different world. As Ian Hacking puts it in his comment on Coetzee's story with the three hens, the impact of the experience is not to do with knowledge, with gaining new information; it is more basic than that, more existential.

Granted that the two experiences at hand share this general structure, the "conversion" they bring about is not at all the same. Coetzee, his readers are invited to think, could not go on eating animals such as hens after the experience. Plumwood, on the other hand, did not give up eating animals, or supporting certain practices of hunting, "after the crocodile". Indeed, two instances of conversion, with very different consequences as to the kind of change they brought for their subjects.

How to explain this difference? Perhaps the answer is simply that the animals are so different in the two cases: three hens and a crocodile. If anything, they certainly drive home the point made by several of the philosophers discussed earlier, from Midgley to Derrida: that animals make up a vastly heterogeneous category. To the extent that the two experiences at all raise the same set of questions, say, about whether to hunt and to eat the animals in question, the fact that hens make up the one case and a crocodile the other seems to call for a differentiated answer rather than a one-fits-all one. Obviously, what the crocodile can do to Plumwood – kill her if it (sorry: he or she) decides to – the hens cannot do to the boy in Coetzee's example. Moreover, as Coetzee observer, hens, sheep, and cattle are animals that we humans have "brought into our world", domestic at that, which makes a difference as to our responsibility for how we treat them. By contrast, a crocodile embodies a distinct autonomy, sovereignty, and independence of us humans, spectacularly and majestically so, thus hammering home a very different point than do hens, sheep, and cattle – namely how inferior we are, and so, as part of that, how misplaced and hubristic it would be were we to assume responsibility for such an animal as a crocodile (whatever responsibility would mean here), its being wild, its *not needing to have anything at all to do with us* being, arguably, the hallmark of its existence.

Before we look more closely at the importance of distinguishing between different sorts of animals, I wish to pay attention to a dimension that is merely hinted at in Plumwood's book, yet one I take to enrich her perspective – namely, the dimension of the animal soul.

In *The History of Animals: A Philosophy*, Oxana Timofeeva draws heavily on the work of the Russian writer Andrei Platonov, especially the novels he wrote in the 1920s. Platonov develops the notion of "a poor life" to describe the life of animals and plants, and Timofeeva contrasts it with Heidegger's famous assertion, mentioned when we discussed Derrida, that animals are "poor in world" ("*weltarm*") and that their lack of hands goes to demonstrate how radically different are their lives from that of humans, who alone actively relate to their world. Timofeeva explains that Platonov is "clearly on the side of those small, poor and weak living beings", and he attributes to them "a certain inner virtue of existence" (2018: 154), one denied them not only by Heidegger but by the entire tradition of Western thought he echoes. She quotes this passage from Platonov:

> Surely not every animal and plant could be sad and wretched; this was a dream or pretense of theirs, or some temporary disfigurement they were suffering from. Otherwise one would have to assume that true enthusiasm lies only in the human heart – and such an assumption is worthless and empty, since the blackthorn is imbued with a scent, and the eyes of a tortoise with a thoughtfulness, that signify the great inner worth of their existence, a dignity complete in itself and needing no supplement from the soul of a human being.
>
> (Platonov 2008: 120; Timofeeva 2018: 154)

If anything is poor and makes for impoverishment, then, it is not the inner lives of animals, even small ones without any of the physical prowess of a crocodile, but the self-imposed taboo among humans against acknowledging the qualities of those inner lives for fear of anthropomorphism. Platonov's rebuke to Heidegger et al., then, is to assert that, "yes, animal life is poor in world, but the world itself is even poorer and is unquestionably in need of the energy of this life in order to resist the forces of entropy and death" (Timofeeva 2018: 155). Hence, "if Heideggerian animals cannot give, but only take or grab, then in Platonov only poor life has the ability to give"; poor life "generously shares itself, and animals give to humans their life substance, their 'soul', together with their flesh" (Timofeeva 2018: 156). Accordingly, we find in Platonov a very different portrait of humans' eating meat from that found in Coetzee's work:

> When his father brought birds and animals back with him, he had eaten them thriftily and sensibly, teaching his children to do the same, so that an extinguished gift of nature would be transformed to man's benefit and not go to waste down the latrine. He used to say that the meat and bones of dead creatures should do more than just fill you up; they should also provide you with a good soul, and strength of heart and reason. If you can't take from a bird or an animal its most valuable good, if all you want is a full stomach – then eat vegetables, eat cabbage soup, or bread crumbled in water. His father held that the world's animals and birds were precious souls, and that love for them was sound husbandry.
> (Platonov 2008: 157f.; Timofeeva 2018: 157)

Timofeeva's comment on this passage is worth quoting:

> Animal flesh in Platonov serves for "feeding the soul" and at the same time gives to the body its own "good soul". The human soul is thus eating the animal body, while the human body is eating the animal soul. Platonov's soul, which eats flesh which feeds the body, is material; it has nothing to do with the incorporeal Christian soul, ascending to heaven after death. It is an *anima*, the matter of life, which is one of the names for the sub stance circulating among bodies, from animal to human and between humans themselves.
> (Timofeeva 2018: 158)

Although the life led by animals may be "poor", it is nevertheless a life that has its own virtues, such as diligence – animals are working to maintain their lives – and generosity, as displayed in their ability to give, to make a gift, to be a sort of life that is crucial to the lives of others, including humans. What Platonov provides is an appreciation of how animals are important to the lives of humans in that for "us" to eat "them" is to incorporate their soul, something spiritual in the midst of the material, a sort of existential cross-species exchange by way of securing survival, the perseverance of life, of the *conatus* that Levinas is so wary of, mistakenly at that. I see a strong affinity between Platonov's stance and Plumwood's keen eye for what she calls "the ethical continuity of planetary life". This continuity – ethical, ecological – is precisely what is being lost from view as long as the different species – crudely: humans versus animals – are sought identified and described *separately*, in splendid isolation from each other, thus neglecting the extent to which the different species are what they are *in virtue of what they are for each other*, not apart from each other.

Given this background provided by Platonov, we return to Plumwood's discussion of the issue of whether or not to eat animals for food.

Plumwood faults what she calls "the absolutist vegetarian" for generalizing "the animal food construction of a specific abusive culture to the construction of any possible culture" and considers the work of Carol Adams a case in point, commenting that: "To the extent that the ruthless, reductionistic and hyper-separated treatments of animals as replaceable and tradeable items of property characteristic of the commodity form and of capitalistic economic rationality come to appear as inevitable aspects of animal food and of human predation or consumption – which, of course, they are not – total abstention from animal food can be made to appear the only possible course" (2002: 158). To my mind, this criticism helps illuminate why it is so important to recognize the *richness* of practices where humans and animals interact, a richness lavishly present in the work carried out by historians and anthropologists alike.

This historical as well as cultural richness, across epochs and societies, is not taken into account but is instead conspicuously absent in much of present-day philosophical discussions about "the animal question" understood as a purely, or primarily, moral issue about humans eating animals, where the answer is sought in a once-and-for-all fashion, strictly and rigorously principled and universal at that, and where *one particular* way of treating – with the overall objective of killing and eating – animals is taken as prototypical, excluding all others, namely that of present-day, large-scale factory farming where "humans" and "animals" are not only set apart conceptually and ethically, homogenizing what is truly heterogeneous biologically and ecologically speaking in the case of "animals", but setting them apart *physically* as well, thus entrenching a cultural and economic arrangement where the different species do not meet, do not interact, do not *coexist so as to matter to each other*. As a consequence, the kind of practices Platonov writes about appears exotic, even though they were perfectly common, in most parts of Europe, as recently as two generations ago. As we saw, these are practices allowing for each new generation's cultivation of species-specific virtues on both sides of the now-erected divide: the diligence and generosity of *giving* on the part of those creatures allegedly poor in world, and the frugality, respect, and gratitude of *receiving and responding* on the part of the humans who depend on, and who individually and collectively, by way of seasonal rituals and the like, recognize they depend on that gift.

In this way, philosophical and cultural impoverishment now become two sides of the same coin, each reinforcing the other, foreshadowing what we in the next chapter will examine in the form of the link between ecological extinction and extinction of experience. What happens is that *one* way of answering the question about the morality (or rather immorality) of humans eating animals is tailor-made for *one* sort of economic arrangement as to how the animals in question are treated. This one-to-one constellation between principled stance on the one hand and practical arrangement on the other comes with the price of precluding and suppressing alternative stances as well as alternative arrangements, making for a poverty partly intellectual and cultural, in terms of imagination and experience, and partly utterly concrete and literal.

This is my way of putting it, but I consider it perfectly compatible with Plumwood's position. Shifting her focus from absolutist vegetarianism to "ontological veganism", Plumwood contends that the latter:

> strives to extend the dualised human status to the non-human area, in the name of equality, re-envisaging non-humans also as beyond any form of use. Thus Carol Adams argues against any use of the animal other (for food or anything else) as involving instrumentalising them. Stating that "the ontologizing of animals as edible bodies creates them as instruments of human beings". [...] Instrumentalism is misdefined by Adams as involving any making use of the other, rather than the treatment of the other as *no more than* something of use, as merely a means to an end. We must inevitably treat the natural world to some degree as a means, for example, to food, shelter and other materials we need in order to survive, just as we must treat other people to some degree as means. What is prohibited is unconstrained or total use of others as means, reducing others to means. We cannot give up using one another, but we can give up use/respect dualism, which means working towards ethical, respectful and highly constrained forms of use.
>
> (2002: 159)

To be sure, a strikingly Kantian way to argue a position that breaks with anthropocentrism a la Kant, yet without giving up on a notion of uniquely *human* responsibility for ensuring that *no one* – be it animals or fellow humans – be reduced to mere means. For, to repeat, there *is* something uniquely human, and it is precisely to do with an obligation that we have to respond responsively to creatures, such as animals of different sorts, who, to varying degrees and in various forms, *share* our vulnerability and our dependence on others, yet themselves *lack* the capacities required to reciprocate our responsibility for them, this being the essence of the asymmetry between agents and addressees that I have been talking about throughout.

Ultimately, the poverty of the moral "absolutism" that Plumwood argues against can be said to consist in the inability to make the distinction between *use* and *abuse* (of animals by humans): *any use* is tantamount to abuse, to be considered morally suspect, morally wrong. This connects with one of my major points when I used Winnicott to criticize Levinas: the subject must be allowed to use the object, use it by way of testing it in practice so as to experience its realness, its powers of resistance in being not me, different from me, a site of agency separate from me and within reach of my powers, yet for all that beyond my control.

Admittedly, this subject-object way of putting it is abstract and technical, in keeping with the terminology of psychoanalytic theory that Winnicott's work employs. Its relevance with regard to Levinas, however, is that participation in a moral relationship with another cannot take the form of disallowing oneself to come close enough to the other to use him or her or it (if "it" per convention stands for the animal other); cannot, that is, make "do not touch, stay at a distance" the required stance of the agent. Rather, to apply Plumwood's distinction, *respect for the other goes via use of the other*, use that can be – and that is the point – either morally appropriate or not. Moral performance, right or wrong, for good or bad, is situated *within* that testing-using engagement with the other, acutely attentive to his or her or its response – unfolding within the messy and imperfect, frustratingly so, but situated there nonetheless. There is no other arena where morally relevant choices can be made; they can only be made where they are enacted and where their consequences will play out, sometimes staying there, limitedly, locally, sometimes being carried far beyond the spatio-temporal particularity of the arena where agents and addressees meet.

Having said this, and as noted in my discussion of Habermas, we must keep in mind that under present-day conditions those affected by a phenomenon such as climate change will often be invisible, unknown, unencountered others, faceless instead of distinct faces, at the receiving end of actions that for their part multiply and intermingle as far as their "first movers" in the sense of distinct (individual or collective) agents are concerned, rendering the model where an agent enacts his or her responsibility for his or her "other" within a setting of shared spatio-temporal presence – hence "encounter" – at best a heuristic device, an "ideal type" in Max Weber's sense and as such often not matching the actual circumstances where the consequences of actions travel across continents and far into the future.

IV

I mentioned that philosophers' discussion of whether it can be justifiable for humans to kill animals is often faulted for failing to distinguish between different sorts of animals, and rightly so. It goes without saying that acknowledging how much various animals differ from one another is necessary for a host of questions about how humans should relate to them besides that of whether or not to kill them.

Plumwood approaches this topic by observing that the dissociation of use and respect now "underlies the uncomfortable division in contemporary culture between the pet animal and the economic animal" (2002: 160).

The use/respect dualism is one of a long-standing and culturally entrenched series including reason/emotion, public/private and person/property, combining to construct the present-day contrast between the private pets – such as the cat who looks at a naked Derrida in the bathroom – and the faceless "factory-farmed" animals, whereby the pet animal is only partially commodified whereas the "economic" animal is fully so. Constructed as hyper-separated and complementary animal categories, the pet enjoys the privilege of emotional attachment of its owners, though of course variably so, whereas the economic animals are undersubjectivized and emotionally divested such as to be reduced to mere "meat" and so a mere means.

These two sorts of animals represent the top and the bottom of the prevailing hierarchy. As we saw when discussing Shepard in Chapter 1, historically the majority of animals have belonged to neither of these two categories. Rather, they have been somewhere "in between" as far as both classification and treatment are concerned – a way of putting it that makes sense only now, retrospectively, however, for lack of having been hegemonic until relatively recently. Keeping animals in the wild out of the picture for now, say, the dogs that lived on farms were neither pets nor economic animals; more outdoors than indoors, they would contribute with their particular skills in helping their "folks" out with protecting some animals (say, sheep) against other animals (say, wolves), being domestic only in a semi-sense that guaranteed them a certain amount of independence from humans.

Do Rose's dingoes fit into this typology? Not quite. Plumwood mentions that in earlier times, the hunting, farming and shepherding man would make a contract with certain wolves, so that in return for a critical task, they would be respected, meaning that "they would never be meat": in helping construct other animals as meat, then, "not only would they themselves never be meat, they would be 'looked after', given a share of the meat themselves". Working animals like sheepdogs would often be a "familiar" animal and as such the subject of deeply personal relationships, helping side them with their human-folk rather than with their companion animals in that the contract is essentially about mutual benefit in meat, thus having "everything to do with complicity" (2002: 161).

Today the "familiar" animal in this role has largely disappeared. Its place has been taken by the polarity of the now underemployed pet animal (more indoors than outdoors) and the fully instrumentalized "meat" animal, where the latter, in increasing, staggering numbers live their entire lives within the super-controlled and -monitored regime of industrial mass production imposed on them from birth to death in the factory farm, the fish feedlot or the laboratory, reduced to biomass for all practical purposes, which is to say as commodities. When the pet and the meat animal come to exhaust the roles available to animals in contemporary society, one of the prices paid is precisely the disappearance of the animal "familiars", the animals that were at one and the same time integrated into our economic activities and our affective lives, taking an active part in *a shared life-form* at that. With this loss, what we are left with is the poverty of the pet/meat binary, foreclosing the possibilities for humans to experience in full display the skills and powers of animals in an outdoors setting, to being-with animals whose subjectivities are appreciated, letting them become "significant others" in the way that only being *neither* a fully domesticated pet *nor* an economic animal allows for.

The shift from what Plumwood terms the old contract to the new one makes for a contradiction that did not exist before, at least not as crudely, and that involves a distinct sort of hypocrisy: "our simultaneous claim to love some animals and to have a right to ruthlessly exploit other animals who are not very different, to simultaneously admit pet subjectivity and ignore or deny meat-animal subjectivity" (2002: 163). Self-declared "animal lovers" honour this dualistic contract when they, sometimes vegetarians or vegans themselves, bring into existence and even breed carnivorous pet animals whom they feed on the "meat" of other animals. The problem goes deeper than such hypocrisy, though. The bottom line is that the contract "cannot be extended to provide liberation for all". Why? Because there is no way of denying that it is essentially an exclusionary contract where "some make a living by complicity in instrumentalizing, imprisoning and oppressing others" (2002: 164). As we saw, not only humans are part of this regime, so are the animals themselves, pitted against each other so that some have to die in order for others to live.

How can we move beyond this contract to enable a relationship between humans and animals not pervaded by contradictions and double standards? Plumwood proposes to start from the concept of the "familiar" animal introduced above, observing that "the relationship with the working animal was often strongly communicative, built-on a respect for animal difference" (2002: 165), bringing back instead of splitting the economic and the emotional connection with the animal. But wouldn't such an animal to a large extent be confined to the conditions of a domestic one? Not necessarily. "If you are very lucky", writes Plumwood, a wild free-living animal in your local surroundings can be a "familiar" animal, provided that you can see it (him or her) often enough to come to know individually and so become a significant other; indeed, there is a contradiction human friendship and the "wildness" of an animal only "if we define 'wild' only in the terms of the absence of the human, rather than as 'free-living'" (ibid.). She mentions relationships with local lizards, birds, and friendly mammals like wombats, but she also includes some of the animals that suffered most under the contract, such as hens, ducks, and geese as participants in "a non-oppressive form of the mixed community" (2002: 166).

There is certainly a very delicate balance involved in this kind of relationship, one easily tipped in one of two directions: either the animal is drawn so closely into the life form of its human counterparts as to lose its integrity inasmuch as it depends on its independence, or it remains so much an animal enjoying its (relative not absolute) freedom from interaction with humans as to be beyond the reach of becoming a "significant other" for particular human persons, persons who for their part hold some special significance for the animal, so that there is a genuine element of attachment and loyalty going both ways. I suspect that Shepard would think that Plumwood's way of striking the balance is too human-centred, as betrayed in the word "friendship", though less so than he would hold for most philosophers these days.

Having distinguished between different sorts of animals, making for different sort of relationships with humans, the issue of their moral status remains. Predictably, Plumwood dismisses the "extensionist" position that allocates moral consideration to non-human beings entirely on the basis of their similarity to the human, so that the greater the similarities, the higher the standing accorded the animals. Instead of being assessed in light of the hegemonic human centre, an independent one "with potential needs, excellences and claims to flourish on their own" should be established (2002: 167). But what precisely would it mean to discard similarity to the human as the basis for moral worth?

For a start, it would mean to "decenter the human and break down human/nature dualism on the ethical front"; that is to say, to frame the issue in terms of "studying up" rather than "studying down", i.e., looking for the source of the problem in the oppressor rather than the oppressed. This means that "it is not so much a question of whether earth others are good enough for ethically rich relationships, but of whether we (western) humans are" (2002: 168). I shall not review all the elements Plumwood lists as part of her proposal for a "counter-hegemonic" framework, only observe that avoidance of unnecessary species-ranking and making the human/other species distinction less central to our moral thinking are among them. She is adamant about what sort of approach has come to an end and reached a stalemate, namely the one where philosophers have stood at a safe distance from the city gates, arguing about the applicability to non-humans of highly abstract ethical concepts like "intrinsic value", without actually going through the gateway and establish specific ethical relationships. Indeed, the concepts framing these debates are no more than "gateway concepts"; the debates they facilitate are "empty to the extent that they evade the real moral task of developing an adequate ethical response to the non-human world, which they do not address in any specific, rich or useful way" (2002: 169).

Plumwood illustrates what it would mean to downplay the human/other species distinction when she suggests that "dealing with a strange highly venomous snake who has just moved onto your veranda may not be all that different from [the ethical challenges involved in] dealing with a difficult human stranger who has done the same" (2002: 170). It is sobering, and very much in Mary Midgley's down-to-earth arch-British spirit, to be invited to reflect on what one would do in such a situation rather in the prisoner dilemma-infected ones so abundant in much of analytic philosophy; after all, it's not that many of us who ever has to decide whom to save from being hit by a trolley or a wheelchair, let alone to justify our decision in front of an imagined jury of professional ethicists afterwards. The questions I would need to answer in Plumwood's example are utterly concrete, yet for all that no less at the core of ethical principles, such as: whose needs should be prioritized, mine or the snake's, or mine or the stranger's? What drove them to my veranda, and are there alternative places to stay that would offer the same protection, say, in the case of a flood or extreme rainfall? How long would their stay last, and how much weight should be accorded to my status as the owner of the place? Would the benefits for the snake or the stranger in being allowed to stay as long as needed outweigh the disadvantages for me? Not out to come up with a particular way of dealing with the uninvited visitor that would be "the" morally right, let alone mandatory one, Plumwood's goal is to establish that "in neither case, *as a matter of justice*, are you entitled to initiate unnecessary and disproportionate violence, shooting the snake, or the difficult human, for example, just on the grounds that they may become a problem" (2002: 171). The main point is that "the same sorts of general ethical approaches can be applied in each case" (ibid.), so that we are not well advised to take *one* approach when other humans are the affected party, and *another* when animals are. Rather, we should recognize that sometimes humans and animals do not differ very much but are pretty alike as far as their needs and our corresponding responsibilities are concerned.

This reminds us that the mere knowledge that the "other" is an animal, not a human, is not in itself sufficient to answer the question about what to do. What sort of animal it is matters; and in cases where we welcome a particular animal into a shared lifeworld by, say, letting the snake in Plumwood's example stay on our veranda, the longer such an animal stays in a place permitting daily encounters, the more we will come to appreciate the individuality of that particular animal – as will it (he or she) as well with regard to us humans. Genuine recognition requires appreciation of the individuality of the other, regardless of whether the agent or addressee is animal or human; to be worth its salt, recognition cannot simply – crudely – refer to species-membership but must affirm as valuable the flourishing of *this particular animal's* individuality. Pet owners do this all the time, endlessly comparing the newly acquired cat or dog with its predecessors in the family. There is no reason to restrict this element of recognition to pets, however, even though of course in most circumstances it will be pets that we come to know the best. For all that, horses, hens, and pigs are no less truly individual beings, fully capable of distinguishing between their human counterparts, based on good experiences with some of them, less so with others. That animals in the wild such as wolves are true individuals needs no argument here.

Plumwood writes perceptively about the importance of our willingness and ability to recognize the non-human other's (be it a particular animal's) potential for intentionality, for communicative exchange and agency; our openness to potentially rich forms of interaction and relationship will also have an ethical dimension, bearing on who we "invite in". To think of this as a purely individual decision would however be to misconstrue it. From early childhood, the culture we belong to instils in us the "correct" way of distinguishing between appropriate and inappropriate others, as we saw in Chapter 1. Thus, to illustrate, at the age of three my son used to greet the trees and the creek when spending his Sundays in the forest, but at the age of six he had stopped this animistic practice of his. Why? He came to view it as childish. It was evidently out of tune with how grown-ups like his father acted, restricting greeting to the exclusive category of fellow humans, excluding all other others encountered in the forest, be it a birch tree, a dove, or a hare. For a six-year-old child in present-day Western society to stop greeting non-human beings of any sort met upon in the forest is evidence of successful socialization. It is something that happens as a matter of course, so that the burden of proof is placed firmly on those who would question or actively deviate from what is sanctioned as "correct" distinction-making. Every culture naturalizes its impact on its individual members by way of socialization, erasing the traces of its impact in the very act of enacting it on behalf of the community, so that the high degree of *selectivity* entailed is effectively hidden from view, depending as it does on comparison with other cultures, or epochs, to become fully conscious to the individual member. That things could be otherwise, that various others can be looked upon and related to in profoundly different ways from the one internalized since childhood, is the great counter-factual not considered. *Pace* Habermas and other grand theorists of Western modernity in the wake of Max Weber, this holds not only for so-called "mythical" (pre-modern) cultures but for the hegemonic modern Western one as well; if anything, the homogenizing pressures when it comes to demanding adherence to a specific cosmology – one where human exceptionalism is a key feature – can hardly be overestimated, though there are signs that it faces new challenges with the advent of the Anthropocene (I return to this below).

Philosophers too often neglect the cultural dimension I highlight here. Such neglect is a result of their predilection for focusing in abstract, "principled" fashion on which (species-specific) capacities it is most appropriate to regard as the relevant ones in determining the status of the other in question: is it a non-human other entitled to our inviting it in, morally speaking, or is it not? In the many cases where we decline that gesture of "inviting in", we usually do so (as Plumwood observes in her criticism of Daniel Dennett) "for reasons that have little to do with differences in animal minds and abilities and a lot to do with our own choices about which others to subsume under an instrumental and reductive rationality in order to free ourselves from ethical constraints in our treatment of them" (2002: 185); choices that must be understood as reflecting the cultural impact I spoke about and not as choices up to us to make in a strictly individual manner.

To be sure, there is a sense in which you may *try* to break ranks and decide fully on your own, autonomously, about whom to grant full moral standing and whom not to, say, by way of relating to a particular squirrel as no less a "significant other" than members of your own family. But know that sanctions will be quick in coming once that "freedom of choice" is identified as the deviation from the cultural norm it truly is: if intended as a cultural protest, it will most likely not be responded to as such but instead individualized – "return to sender"-like – in a manner reflecting on the idiosyncrasy exposed in the person, psychopathologically understood and so calling for psychiatric treatment (should deviant behaviour persist) rather than seeing the deviation as a possibly well-grounded corrective to dominant cultural norms.

Thus the problem is not simply that our culture, as assessed from the critical point of view of ecophilosophers like Plumwood, is one that draws the distinction between those others let "in" and those left "out" in a questionable manner. It is also that individuals who criticize it as doing so, and who enact that critique, demonstrating alternative ways of relating to, say, certain animals, are themselves subject to a shift from being inside to being placed outside. *Excommunication* works in several directions and may target individuals on both sides of the human/animals divide posited and monitored by the culture one belongs to, challenging the self-serving notion that ours is a more tolerant culture than that of other societies or previous epochs.

V

In a number of books, most notably *Animals and the Moral Community* and *Animals and the Limits of Postmodernism*, the American philosopher Gary Steiner discusses the issue of animal rights. He does so in a more rigorous manner than Plumwood, sharing many of her views and concerns. Nevertheless, the position advocated by Steiner differs significantly from Plumwood's in that he regards "universal veganism as a strict ethical duty" (2013: 198), arguing that "what is absolutely clear is that cosmic justice demands universal veganism, the refusal to consume animal products of any kind", adding that "the more inured we are to anthropocentric values, the more unreasonable, burdensome, and impracticable this proposal will seem" (2008: 163).

First, we need to understand how Steiner's position differs from Plumwood's, as well as from Rose's and Shepard's. All three of them hold that, as Rose puts it, "the gift of life, like the face of death, comes wrapped in blood". They all support indigenous hunting, and also certain forms of recreational hunting; indeed, Rose and Shepard would participate themselves. We also know that they oppose factory farming on the grounds that within this regime the entire life cycle of the animals is controlled and organized with one objective in mind: to reduce the animals to the mere means of being mere meat, maximizing their being-nothing-but-meat for the solely human-centred purposes of consumption and profit. Thus for Plumwood,

Rose, and Shepard the question about killing animals does not allow for *one* answer, one that would hold "always and in all instances" in the heterogeneous manifold of cases where killing, in some form or another, is an issue. Rather, the answer is that of multiple "it depends": on the context, on what sort of animals are affected, what sort of methods are used and attitudes are displayed on the part of the humans, for what purposes, and with what consequences for the larger biotic community the animals are part of. The bottom line, where the ecological and the ethical are two sides of the same coin, is that "all ecologically embodied beings are food for some other beings" (Plumwood 2002: 155). Hence the three ecophilosophers find ecologically uninformed and ethically misguided the notion so widespread in so-called "progressive" philosophical circles, as well as in influential versions of "environmental humanities", that "predation is fallen", to recall Plumwood's formulation (2002: 155).

Why is it, then, that Steiner arrives at a different conclusion, advocating universal veganism as a strict ethical duty, implying a refusal to consume, for human purposes, animal products of any kind?

Steiner's point of departure is similar to Plumwood's: like her, he rejects the conventional assumption that superior cognitive abilities are the basis for superior moral status; that is, he rejects that moral status is in any way a function of cognitive capacities (2008: xi). This means that he dismisses appeals to what Gary Francione calls "similar minds theory", where "an animal's moral status depends on the degree to which that animal's mental life is sufficiently like the mental life of human beings" (2008: 90). On Steiner's view, "the cognitive differences between humans and animals have no moral significance whatsoever, just as more highly intelligent humans do not enjoy greater moral entitlements than less intelligent ones" (2008: 129); in saying this, Steiner implicitly alludes to a well-known tension in Kant's position on moral status among humans. Even though Steiner deems it important that we rethink our view about the cognitive capacities of animals, allowing them more advanced such than does the traditional view, his fundamental point is that such capacities, having their "model" in humans, not be taken as the standard for determining the moral status of animals.

It turns out, however, that Steiner himself does not do without invoking capacities altogether: while consistently rejecting that *humans* be considered the standard against which animals must be measured to be granted moral worth (i.e., the greater their similarities with us, the greater their worth), he does retain a role for capacities, namely in arguing that "sentience – the capacity to experience pleasure and pain – is a sufficient condition for equal consideration in the moral community", where sentience is understood by Steiner as "a capacity shared by all beings for whom the struggle for life and flourishing *matters*, whether or not the being in question has a reflective sense of which things matter or how they matter" (2008: xi). The important question to ask, then, is which capacities are and are not relevant. Based on the primacy he grants sentience in this respect, Steiner proceeds to use the language of animal rights to argue that "humans have basic obligations never to use animals for food, clothing, entertainment, or experimentation, even though animals have no corresponding obligations – indeed, no moral obligations whatsoever – toward humans" (2008: 92).

In keeping with this view, Steiner endorses what I have argued is the fundamental asymmetry between humans and animals, morally speaking. He puts it like this: "This asymmetry consists in the fact that it is possible – and, I argue, morally incumbent upon human beings – to recognize that we have fundamental obligations toward animals…in spite of the fact that animals are fundamentally incapable of taking on reciprocal obligations toward us" (2013: 177). On this view, "nonhuman animals are incapable of being moral agents" (ibid.), the point being, as I have put it throughout, that animals do not need to be moral agents to be regarded as moral addressees. Steiner's way of stating this is to say that we have duties towards animals that for their part have rights, yet without the animals having corresponding duties towards us. As we have seen, I express the asymmetry in a way that does without the concepts of rights and duties; instead, I submit that we humans have responsibilities towards animals even though they have no responsibilities towards us. That is the crux of the asymmetry.

A large portion of Steiner's *Animals and the Limits of Postmodernism* is devoted to a critical assessment of Derrida's views about animals. On the one hand, Steiner agrees with Derrida's contention that Heidegger "effectively belittles animals by reducing their deaths to the status of merely passing out of existence" and so fails to do justice to the moral gravity involved in the death of an animal (2013: 178). I want add that what complicates the assessment of Heidegger's position is that he pays no systematic attention to what sort of death "the animal" is thought to experience, its being killed by humans being but one possible death among many. On the other hand, Steiner considers it "ironic" that Derrida himself "avoids making any definitive claims about the right of animals not to be eaten by human beings" (ibid.). It does not help that Derrida once declared that "I am a vegetarian in my soul" when in fact he "was neither a vegan nor a vegetarian", as Steiner reports. Looking in vain for a statement in Derrida to the effect that "killing animals is wrong, period", Steiner concludes that for Derrida, "any definitive moral claim about anything at all would bring the activity of questioning, which is essential for authentic responsibility, to a standstill" (ibid.). In advancing this criticism, Steiner seizes upon Derrida's statement that he is "not recalling [the violence that traditional philosophy has done to animals] in order to start a support group for vegetarianism, ecologism, or for the societies for the protection of animals" (2013: 126). When everything is said and done, then, for Derrida as Steiner reads him calls for vegetarianism may be well intended, but they "nonetheless are impossible to implement and simply reproduce violence", be it the symbolic violence of the Western philosophical canon. Steiner's comment is damning: "the abdication of responsibility in these ideas is breathtaking" (ibid.).

To be sure, and as we saw in Chapter 3, there are good reasons for reacting with exasperation to the Levinas-inspired formulations where Derrida insists that responsibility is "without limits", "excessive", "before memory", and "exceeding all calculation", mystifying the phenomenon of reality by upping the rhetorical ante to such a degree that no agent can ever be sure about what, concretely, his or her responsibility consists of in a given situation, let alone what actions it demands. Steiner raises the question whether "deconstruction [a la Derrida] leads to irresponsibility" and indicates an answer in the affirmative. This being so, he endorses Paola Cavalieri's claim that "Derrida simply erases the problem of the value of animal life by dismissing philosophical vegetarianism" (2013: 127; see Cavalieri 2009: 98f.).

Much as I share Steiner's exasperation at Derrida's rhetoric, I find the argument he takes over from Cavalieri to be a category mistake, and so unconvincing. To hold that not supporting "philosophical vegetarianism" equals erasing the problem of the value of animal life is not to make a compelling argument. Instead it reveals an attitude of philosophical intolerance towards all positions that deviate from one's own, i.e., that the moral question concerning the killing of animals by humans has one and only one valid answer, holding across the spectrum of cases, methods, purposes, etc. mentioned above, namely that such killing is wrong and must be prohibited.

But Steiner's objections to Derrida only take us so far when it comes to recognizing how his own position differs from that of Plumwood, Rose, and Shepard. In this regard, his reference to Aldo Leopold is instructive, taking Leopold to task for his advocacy of hunting and his acceptance of humans' eating the meat from animals. The question is: is condoning killing animals compatible with truly respecting them? Or does the one exclude the other?

Steiner engages with the work of James Hatley who, like Leopold before him, holds that respect for animals and involvement – say, by hunting – in the cycles of predation are inseparable from one another, and so not at all incompatible. Hatley argues that Western culture's premium on humans' alleged superiority in cognitive capacities has led us to forget that we are part of nature. The possibility of being the prey of a wild animal – they very thing that happened to Plumwood – is a vital reminder of what is suppressed: "Hardly anything", writes Hatley, "could be more intimate than becoming the food, and so the body, of another animal in the wild" (Hatley 2004: 14f.; see Steiner 2013: 216). To take seriously that we can ourselves be the prey of some predatory animal, then, gives the lie to the popular notion that we exist outside the context of predatory activity. As a result of historical development, however, our appropriation of nature has enabled us to "make over the space in which we live *as if* humans had become inedible and everything else is revealed to be more or less available for ingestion" (ibid.). This situation helps sustain an asymmetry where we are predators and never prey, upholding the notion that we are in some profound way different from other natural beings, including animals. We thereby, contends Hatley, fail to acknowledge that "we only get to be a human, or a bear, or a microbe by eating others" (ibid.). To attain a more adequate relationship to the natural world, Hatley suggests, we need to recognize that we are immersed in that world yet also at a distance from it, owing to our unique capacities for reflection. Yet Steiner faults Hatley for focusing more on "our making ourselves available to predatory animals" than on animals' making themselves available to us. It is difficult to see how this model will work out in practice, however. How, more precisely, are we supposed to offer ourselves, "at least symbolically", to predatory animals as their sustenance? Can there be, say, an obligation to make oneself "available" to an animal like the crocodile that Plumwood encountered in such a dramatic way?

Since Steiner offers no answer to this question, I suspect that his reason for discussing Hatley's model is to show that an ethics based on the premise that human and animal life alike is about "eating and being eaten" is essentially mistaken: it suggests a kind of reciprocity between the species that cannot be enacted. It is simply not plausible that human persons "offer themselves" to predatory animals in the setting of contemporary society. That this is indeed Steiner's view becomes clear in his discussion of the French philosopher Dominique Lestel's provocative thesis that meat eating is not only permissible but incumbent upon us. Lestel's "ethic of carnivorism" champions the ability to receive as just as important as giving, citing indigenous peoples who thank their animal prey's spirit for the "gift" it has given, and helping us see that predation should not be considered as aggression but instead as a principle that governs the energies that circulate in the world. On Lestel's view, then, eating animals is not simply a biological necessity but helps sustain a cosmic human/animal relationship of mutual dependence. The upshot is not to declare open season on animals but to instil in humans an obligation to recognize the animality that we share with them and to recognize their subjectivity. In this spirit, the "ethical carnivore" is not one to take meat eating lightly but instead shows appreciation of the "infinite debt" that we humans owe to animals; rituals following hunting see to it that we "celebrate our own animality and the sacrifice that the animal that we eat has made on our behalf" (2013: 219). In keeping with its spiritual significance, Lestel urges modesty in the amount of meat we eat, but warns against stopping to consume meat altogether, since "a life without [some] violence toward animals would lose something that is essential to a life truly worth having been lived" (ibid.).

We have seen that the notion of a shared animality between humans and animals is a controversial one that has met strong opposition from major philosophers. Thinkers as different as Plato, Descartes and Kant have contributed to the hyper-separation (Plumwood) between humans and animals so central to Western culture by suppressing the animality in humans, underpinning the long-lasting resistance to Darwin that Midgley noted in Chapter 1. We also saw that Adorno tried to oppose the destructiveness of this suppression of "nature in the [human] subject", where animality is rendered as something foreign, thus alienating humans from animals as their "radical others". Yet Adorno did not ask the question that Steiner's discussion of Hatley and Lestel revolves around, namely: What bearing does the shared animality of humans and animals, understood as the fact that they can both eat and be eaten, have on the question of whether it is ever justified for humans to kill animals? Does one rather than the other of the two possible answers follow, so that *either* "can both eat and be eaten" points to the ethic of carnivorism that Lestel advocates *or* it points to the sort of "ethical vegetarianism" that Steiner defends?

A central claim of Lestel's is that proponents of vegetarianism tend to forget our animality and our continuity with all living beings, i.e., what Plumwood refers to as "the ethical continuity of planetary life". Whether intentionally or not, they reassert the "thesis of human exceptionalism", mistakenly believing that "human beings can place themselves above the evolution of all other species" inasmuch as the difference between humans and animals is accorded primacy over their similarities: humans, then, are the species – the one and only species – that deliberately chooses not to kill for the purpose of eating. On these grounds, Lestel (and here I rely on Steiner's presentation) holds that the truly consistent ethical vegetarian is "profoundly hostile to life in seeking to eliminate cruelty from the world, inasmuch as the sole means for satisfying [the ethical vegetarian's pacifistic aspiration] would be to suppress all animal life on the earth" (2013: 220). What such a position amounts to, however, is just the sort of denial of life that Nietzsche so scathingly condemned as a typical "modern-day" sickness. The vegetarian's mistake is to take for granted that animals, just like humans, have an interest in not being killed. Yet animals themselves, maintains Lestel, do not consider this an ethical scandal; only humans may do so. Hence the love vegetarians purport for animals comes at the price of idealizing them to a degree robbing them of their very animality.

Lestel also challenge the vegetarian's categorical distinction between animals and plants, extending Derrida's questioning of the traditional response-reaction distinction by suggesting that not only animals but also plants actively respond to their environment. Steiner counters this by saying that "the overwhelming weight of scientific evidence casts tremendous doubt" on the possibility that plants exhibit sentience, citing their lack of a central nervous system, considered a requirement for the capacity for sentience (2013: 221). Thus, on Steiner's view, we have no scientific grounds for talking about "the intelligence of plants" as does Lestel, among many others these days.

To be sure, the acknowledgement that the dividing line between plants and animals is not absolutely clear goes back at least to Aristotle. Even so, Steiner urges that we do not dismiss the response-reaction distinction altogether. So, while we should recognize that humans sometimes "merely react" and animals are capable of responding, there is no basis for holding that the latter applies to plants as well. However that may be, Steiner's focus, like my own, is on the relationship between humans and animals, not plants. He expresses sympathy with Lestel's concern (shared with Leopold and other thinkers in the deep ecology tradition) that "remembering" our animality qua humans is key to highlighting our essential continuity with the entire living world. In Lestel's case, this leads to the claim that "we *must* eat meat in order to engage in this primordial act of recollection" (Steiner 2013: 223) – this being Steiner's way of putting it, not Lestel's.

But, Steiner asks, "*why* must an acknowledgement of our essential animality involve the consumption of meat, ritual though it may be", and so allowing for the "spiritual" dimension that we have seen Lestel refer to? True, he does refer to the carnivorism practised by some indigenous peoples, such as the Cree and the Sioux. Steiner counters by asking "exactly what relevance the cosmologies of these cultures have to our own endeavor to ground our ethical commitments in cosmic holism" – *must* such a position make a place for killing and eating animals (2013: 224)? What if the situation of the kind of cultures Lestel refers to was such that they *had* to consume animals in order to survive? What if our own situation, by contrast, is such that we simply don't need to engage in such practices? And finally, is it not possible to restore to us "a sense of our immersion in and obligations toward nature" *without* killing animals?

The question about what "we" – presumably present-day, urban, Western individuals and communities – may learn from indigenous cultures with regard to how to treat other-than-human beings is complex. I argued as much in *Cosmologies of the Anthropocene*, drawing extensively on the ethnographic research on a number of (more or less) still intact animistic cultures carried out by anthropologists. On my view, there is no simple answer to be found for "us" within these cultures: the differences in circumstances are too big for simple answers, including ethical ones. So it is not a matter of "us" imitating or copying "them" within our urbanized societies, but of maintaining an attitude of openness to the possibility that there are things to be learned, local knowledge and practices to look to for inspiration, besides helping protect the continued existence of non-Western cosmologies wherever they exist, threatened as they increasingly are with cultural extinction, often intimately linked to biological such, the two sorts of losses being intertwined. What we can say is that with few exceptions, the practices of killing animals in which these societies historically, and to a certain extent to the present day, have been involved, say, in countries like Brazil or Ecuador, have been by and large ecologically sustainable; they have, that is, been part of an arrangement between humans and their various nonhuman "others" where all parties have been allowed the place necessary to flourish and to persist, as so many "ecological selves", through the generations and centuries (see Callicott 1994; Kohn 2013).

Importantly, this holds for a situation where the methods and weapons used for hunting, as well as fishing, have remained "low tech" as measured by modern standards, comparably primitive at that. The question whether to "take out" more than might prove sustainable only becomes a *practically urgent* one (as distinct from a philosophically intriguing one) when the technology is so advanced as to make that "more" a realistic option. Only in that case will an ethical, or cosmological, argument kick in, to the effect that "yes, we would be able to take out more than we have used to do, but we decide not to". Here, then, a distinction is made between the (technologically) *possible* and the *desirable*, where the latter is ecologically and ethically informed, a synthesis of the two in the way I have set out in earlier chapters, materializing in a decision aspiring to prove *prudent* in the long run, that is to say, seven generations into the future according to the cosmologies in question (see Kopenawa and Albert 2013). In short, technological-practical capacity does not posit the limit to what is taken out (wild animals included); ecologically informed, centuries-long *in situ* experience with the way the human-animal relationship plays out, does. The role played by animistic cosmology is such as to discourage efforts to "improve" the methods and equipment at hand with a view to making them as extensive, as expansive and in that sense as "efficient" as possible. Not only is such improvement not considered an end in itself. It is not even regarded as a desirable means to the ends that matter most within the "ecology of selves" involved.

Unfortunately, such an ethnographically informed perspective is absent from Steiner's discussion. Its absence renders his argument abstract in a way ill-suited to the topic inasmuch as the practices where animals are killed and eaten by humans come in many historical forms and with diverse ethical challenges and justifications. Steiner does no justice, philosophical or otherwise, to the richness of this material when he wants to reduce the topic to one simple question, meant to allow for one particular answer, namely: should practices of killing animals continue if they are not (or no longer) necessary for survival?

The question may appear precise and concrete. But it's not. To illustrate why I consider it abstract, consider the story philosopher Jonathan Lear tells in his book *Radical Hope*. Plenty Coups, the last great chief of the North American Crow nation, is known to have said that, after the buffalo went away and the hunting, even the stealing of horses, stopped, "the hearts of my people fell to the ground, and they could not lift them up again. After this nothing happened" (Lear 2006: 2). When all the beaver and the buffalo had been killed, the horses taken away, and there being no more Sioux to fight against, the whole project and identity aimed at excelling as a Crow evaporated before their eyes: everything that had used to matter to them in their existence as Crow ceased to do so; henceforth it was *as though* nothing happened. With the disappearance of the animals and all the practices, skills, and knowledge that their presence had helped sustain to such an extent that the animals were the *sine qua non* of being a Crow, of being able to excel in the way of life that is the distinctly Crow one, that whole way of life was left in tatters, resulting in depression as brought on by loss of meaning, purpose, and identity. Lear quotes anthropologist Marshall Sahlins' observation that every culture is "a gamble played with nature" in that it depends on the continued availability of so many other-than-human others (Lear 2006: 25).

The Crow example makes for an illuminating contrast with Steiner's approach. Steiner proceeds as if the question of eating animals can be posited in a sort of principal purity, uncontaminated by the life form within which the humans for whom the question arises live their lives. In Steiner's model, there is no real encounter, no interaction between particular human individuals and particular animals in a historical-cultural setting ripe with particular features distinguishing it from others. Is this a deficiency weakening his argument?

I think so. Steiner is representative of philosophers' predilection for arriving at an unequivocal answer to principal questions about right and wrong. And indeed, there are cases to show that such an answer may be arrived at, enjoying unanimous consensus – e.g., it is always morally wrong for a human being to eat another human being. True, cannibalism is known to be practised in some cultures, but its occurrence is extremely rare, allowing us to talk of a universal taboo. As Shepard in particular has shown, historically relations between humans and animals have not been like that: that humans kill (some sorts of) animals in order to eat them has been the normal to this day.

Yet if past history has been like that, the future may be different. Steiner clearly thinks that the reasons for eating animals used in the past no longer hold. The bottom line is that we can now do without eating animals; going on eating them is simply not necessary in order to survive. But is this simply an issue about survival? Does sheer necessity exhaust the issue as a moral one, does it dictate the (one right) answer that Steiner is after?

I think that the Crow example (and there are many others) goes to show that Steiner's approach is too narrow-minded: it is reductionist and instrumentalist. There is so much more than sheer survival involved in the way the Crow used to relate to the buffalo they hunted. Plenty Coup urges us to understand "how the animals made the Crow human", to use Shepard's phrase. Especially poignant here is the sense that the indispensable role animals played in the making of the individual Crow, and in helping sustain their identity, their skills and their knowledge qua Crow as a group, is only recognized in its full significance the moment the buffalo are disappearing. It was not the arrows of the Crow that caused their disappearance, but the guns of the white people taking over their traditional land and forcing them into the reservation where and after which "nothing happened" (see Netz 2004: 73ff.). Centuries before the word "sustainability" was coined, by European academics and politicians at that, the Crow ensured that no individual would ever take out more buffalo than sustaining them would allow – that is to say, than sustaining the *relationship* between the buffalo and the Crow (and their horses) would allow, stressing the balance between them as precarious and as an ongoing challenge, demanding frugality and prudence in negotiating it with a view to its intactness for future generations of *animals and humans alike*, there being no morally relevant difference between the two species' interests here, equally entitled as they are to a future allowing them to thrive, and to do so together. Indeed, for the one to thrive, the other must as well. That is the essence of inter-dependence, a principle of ecology that ethics needs to recognize as its basis, its grounding, in the conditions of life.

To return to Steiner's case for ethical veganism and its imperative never to eat animals, he wants to found it on "the recognition that our capacity for reflection, and to articulate and strive to live in accordance with principles, imposes burdens on us that no other sentient being appears to be capable of assuming" (2013: 225). The imperative acknowledges Schopenhauer's endeavour to reduce suffering in the world, while recognizing that "suffering is the basic condition of existence" (ibid.). Here I agree with Steiner, only I would add that suffering is such a condition because of the way it springs from the constellation dependence-vulnerability that I have elucidated above, a constellation-qua-condition shared by humans and animals alike, pointing to their shared mortality. He goes on, however, to reject Lestel's notion of an "infinite debt" to animals, arguing that "rather than capitulating to the inevitability of suffering and predation in the world, the ethical vegan takes the position that we can avoid doing so". He then says: "Reducing all life to the amorphous category of 'flesh' is fundamentally incompatible with such an expression of moral concern for animals, as is the proposition that human life is somehow diminished by the aspiration to eschew all violence toward animals" (ibid.).

I am not sure who in this debate is guilty of "reducing all life to the amorphous category of 'flesh'". As to what may diminish human life, the Crow who no longer could be in a relationship with animals involving killing some of them for various purposes, including eating the meat, is surely a case in point, contrary to what Steiner would have us think. His reductionism comes to the fore: he reduces, as the position he argues against does not, the entire human-animals relationship to one aspect, that of killing for the sake of eating. It is as if the only portion of the animals' life cycle that counts is to do with their *manner of death*, being made into the one issue around which the entire ethical debate revolves and with reference to which the repertoire of available philosophical positions is assessed.

Of course, it makes a difference for Steiner whether animals are mistreated prior to their death, as are, say, chicken produced in factories for the sole purpose of optimalizing their commodity value on the market; chicken made to grow (faster, bigger) by artificial means, becoming so heavy as to be unable to move around. As is well known, this scenario is Singer's point of departure in *Animal Liberation*.

I have no doubt that the Crow would find modern-day industrialized chicken-production just as horrible as do Steiner, Singer, and Coetzee, and I readily add myself to the list. What this consensus shows, however, is that Steiner's principled, culture- and context-abstracting, one issue, one aspect oriented approach is unsuited to the task he sets himself: that of answering, once and for all, the question whether it can ever be justified for humans to kill animals. His proposed answer – no, it cannot – fails to be compelling because the way he frames the question is both too narrow and too crude. It is too narrow in that it precludes from consideration the particular context in which killing in some form or other takes place, neglecting the impact of culture and cosmology and failing to address the connection between ecology and ethics where issues of sustainability and prudence are key. And it is too crude in that it induces us to approach the issue in a homogenizing, simplifying manner failing to do justice to the differences between animals. If Steiner's claim is that what the Crow do to the buffalo is just as wrong as what the present-day transnational capitalist agroindustry does to the chicken – since in both cases, animals are being killed to be eaten, and that similarity suffices to show that the two cases are equally wrong and condemnable – then the price he pays is unacceptably high: just as the buffalo and the chicken are viewed as identical cases (both are being killed), so are the humans deemed responsible, the Crow and the agroindustry bosses. Certainly, this two-set identification flies in the face of the all-too-real differences between the two cases: the differences in the lives led by the animals, and the differences in ways of relating to and being-with the animals on the part of the two groups of human agents. The differences boil down to Buber's distinction: an I-Thou relationship in the one case, an I-It relationship in the other.

I want to end my discussion of Steiner by taking up again the ontological perspective so vital to the ecophilosophers we have looked at, and so neglected in the approach of a contemporary American ethicist like Steiner.

Freya Mathews, a prominent member of the group of Australian ecophilosophers counting Rose and Plumwood, makes a distinction between axiological and non-axiological societies that I find instructive. The sort of animal ethics that Steiner represents concentrates on hierarchies (higher and lower) and binaries (ends versus means, intrinsic versus instrumental value) and on *individuals* as the basic entity with regard to which the issue of moral standing, of rights and value, arises. But the individual is not only *morally* paramount; the individual is also the *ontologically* basic entity, one whose links with, connections and commitments to entities-larger-than-itself are viewed as largely optional, to be chosen and negotiated by the individual inasmuch as membership in extra-individual entities is not *essential* to the very being of that individual – rather, group-membership may well prove a threat to it and must for that reason be voluntary not enforced, as thinkers from Hobbes to Nietzsche, Sartre, and Nozick have argued, covering the entire political-philosophical spectrum and demonstrating the persistent support such individualism enjoys (see Taylor 1989).

We will see in the next chapter that this ontological individualism plays a crucial role in the debates between animal rights ethics on the one hand and conservation ethics on the other: where the former is individual (exemplar) oriented, the latter is species oriented. What Mathews helps us appreciate is the importance of tracing the ontological origins of the two types of ethics so as to get at the deeper reasons for their reaching different conclusions, especially on the increasingly urgent issue of species extinction.

The non-axiological view that we may contrast with the axiological one dominant in Western modernity, represented by Steiner, is typically found in indigenous cultures and among hunters and gatherers. In Chapter 2 we saw Old Tim give voice to so-called Tribal or Dreaming Law. The ontology of this Law is one of identifying the patterns in things that enable the "living cosmos" to renew and perpetuate itself in perpetuity. The Law explains how humans and other-than-humans can participate in this pattern, ensuring its vibrancy and coherence. In this function, the Law is at one and the same time concerned with "being" and "doing", with what is and what ought to be, *pace* the separation between facts and values found in axiological thought. Mathews puts it like this: "It is the living cosmos that has given people existence and it details what people *owe* the cosmos in return, what they need – ought – to do to ensure that this generative order is perpetuated" (Mathews 2012: 13), pointing to duties and obligations that transcend individual-oriented compassion.

Recall that the Aboriginal societies Mathews has in mind took their livelihood directly from nature; hunting was not an option, but an integral component of their life form, as for the Crow. In this setting, "what contributes to the stability of the cosmos is not an equality of individuals but of species or kinds" (2012: 14). Thus individuals are inter-substitutable: "it does not matter *which* individuals instantiate the relationships that perpetuate the pattern", the Law being concerned "not to protect individuals qua individuals but to assure the conditions for the perpetuation of a living cosmos". In keeping with this view, the Law is a law of predation; "it prescribes who can eat whom, when and where and under what conditions" (ibid.). As for animals, the Law posits that "animals would care for themselves as part of a living cosmos that cared for itself. Predation was a necessary and integral part of that pattern: humans owed it to the living cosmos to hunt; the species were *sustained* by hunting" (2012: 16).

In current terminology, according to the cosmology expressed in this Law, what is *ecologically required* and what is *morally right* are two sides of the same coin, the same reality, and not in conflict. *Should* the two be in conflict, then because the balance – here: between humans and animals – has been damaged; there is disruption instead of equilibrium, caused by some party's overstepping the limits of their place (niche) within the whole, the living cosmos they are part of and whose thriving as a whole each part depends on. Mathews addresses this situation – all-too-real today – by saying that "now the global ecology is breaking down: we must take responsibility for nature, look after it to the extent that it can no longer look after itself" (2012: 17; see Rose 2000: 229ff.).

To suggest, like I did, that Steiner's position can be identified with the axiological view, is not entirely correct. In *Animals and the Moral Community*, Steiner promotes what he terms "cosmic holism", a view that would appear to have affinity with the (Tribal, Aboriginal, Dreaming) Law whose holism is articulated in its notion of "living cosmos" and the *relational* as distinct from individualist ontology that goes with it. In this regard, Steiner rightly highlights the relevance of Hegel's philosophy, seeking to overcome the individualism Hegel (1991) regards as deeply problematic in contract theory a la Hobbes and Locke, and to a certain extent in Kant's moral and political philosophy. Despite the promise of Hegel's holism, however, Steiner finds that he, like Kant before him, excludes animals from the sphere of reciprocal rights and obligations and so from the moral and political universe, doing so for the standard anthropocentric reason that animals are "nonrational" and "nonlinguistic" beings (2008: 160). This being so, Steiner seeks to "move beyond the realm of *social* justice (i.e., justice among moral and legal agents) and enter the realm of *cosmic* justice, a relation that obtains between moral-legal agents and nonhuman moral-legal patients" (2008: 115). Differences in terminology aside, this is on a par with my own view of moral agents and moral addressees, stressing the asymmetry between them.

Nevertheless, the holism Steiner wants to advance differs from that of the Law Mathews describes, where individuals are viewed as inter-substitutable. Now this notion is considered deeply problematic not only in individual-oriented liberal doctrines but also in animal rights theory. In this vein, Steiner finds fault with the way a "holistic" ecophilosopher like J. Baird Callicott focuses on "biotic community rather than on individuals" (2008: 131). Unfortunately, Steiner does not engage more deeply with the issue. That is surprising, since the objection that holism – generally speaking – is unwilling and unable to respect the inviolability of the individual, owing to its (his or her) *not* being interchangeable or intersubstitutable, is a standard one that any proponent of holism is expected to tackle. In my view, Steiner's holism is shallow rather than deep (to use Arne Næss' famous distinction): it lacks grounding in the cosmological dimension that Mathews highlights, and – even more conspicuously – it lacks input from ecology, thus missing out on the connection between the ecologically required and the morally right that I mentioned above, a connection contested by the critics of holism old and new, arguing that the individual is being – or is not adequately protected against being – "sacrificed" for the sake of the whole (community, ecosystem). (I will say more about this issue in Chapter 5.)

The weaknesses in Steiner's cosmic holism are nowhere more evident than in his concluding statements. He follows Porphyry in holding that "to take on anything like a cosmopolitan standpoint is to affirm our affinities with, rather than our differences from, animals"; Steiner sees this as required to "argue for cosmic rather than merely social justice" (2008: 162f.). But why should – as is clearly presumed here – affirming our *differences* from animals undermines the project of including them in the moral universe, as "patients" or addressees? There is something undialectical and mechanic, indeed anti-holistic and anti-ecological, in juxtaposing affinities and differences the way Steiner does. Aren't their differences from us precisely what we, at long last breaking with anthropocentrism, should learn to appreciate and protect in animals? It is strange that Steiner buys into the notion that affinities with humans *per se* be considered something positive, and differences something negative. I hasten to add, though, that the statement does not sit well with the overall drift of his argument.

Consider, by way of conclusion, what Steiner says in the book's closing passage:

> The goal of cosmic holism as I have articulated it is to reveal the licentiousness of meat eating and other forms of violence toward animals for what they are – injustices that we seek to rationalize by means of specious arguments about the emptiness of animal experience and the moral superiority of human beings. [...] To affirm cosmic justice is to abandon our fantasy of being "the masters and possessors of nature".
>
> (2008: 163)

Do I need to point out that Shepard's *The Others* is conspicuously absent from the bibliography in Steiner's two books? And that this also goes for the ethnographic studies I have drawn upon? To suggest that the peoples and cultures who have historically been most engaged in hunting, killing, and eating (certain sorts of wild) animals – as opposed to imprisoning them within the concrete walls of factories a la present-day chicken production – have subscribed to "the emptiness of animal experience" and thought of themselves as "the masters and possessors of nature" is to get the historical record completely wrong, as my discussion has demonstrated. It is disappointing that Steiner ends his philosophically well-informed work on a note of such gross historical distortion.

VI

I want to end this chapter by turning from theory to practice, looking at a particular encounter between a human person and an individual animal. The human person is the American author Alice Walker, and the animal a horse named Blue. This is how Walker describes their encounter:

It was quite wonderful to pick a few apples, or collect those that had fallen on the ground overnight and patiently hold them up to Blue's large, toothy mouth. I had forgotten the depth of feeling one could see in a horse's eyes.

I was therefore unprepared for the expression in Blue's. Blue was lonely. Blue was horribly lonely and bored. I was not shocked that this should be the case; five acres to tramp by yourself, endlessly, even in the most beautiful of meadows – and his was – cannot provide many interesting events. I was shocked that I had forgotten that human animals and nonhuman animals can communicate quite well; if we are brought up around animals as children we take this for granted. [...]

But then, in our second year, something happened in Blue's life. One morning...I saw another horse, a brown one, at the other end of Blue's field. Blue appeared to be afraid of it, and for several days made no attempt to go near. We went away for a week. When we returned, Blue had decided to make friends and the two horses ambled or galloped along together, and Blue did not come nearly as often to the fence underneath the apple tree.

When he did, bringing his new friend with him, there was a different look in his eyes. A look of independence, of self-possession, of inalienable *horse*-ness. His friend eventually became pregnant. For months and months there was, it seemed to me, a mutual feeling between me and the horses of justice, of peace. I fed apples to them both. The look in Blue's eyes was one of unabashed "this is *it*ness".

It did not, however, last forever. One day, I went out to give Blue some apples. He stood waiting, or so I thought, though not beneath the tree. When I shook the tree and jumped back from the shower of apples, he made no move. I carried some over to him. He managed to half-crunch one. The rest he let fall to the ground. I dreaded looking into his eyes – because I had of course noted that Brown, his partner, had gone – but I did look. [...] The children next door explained that Blue's partner had been "put with him" so that they could mate and she conceive. Since that was accomplished, she had been taken back by her owner, who lived somewhere else.

Will she be back? I asked. They didn't know.

Blue was like a crazed person. Blue *was*, to me, a crazed person. He galloped furiously, as if he were being ridden, around and around his five beautiful acres. He whinnied until he couldn't. He tor at the ground with his hooves. He butted himself against his single shade tree. He looked always and always towards the road down which his partner had gone. And then, occasionally, when he came up for apples, or I took apples to him, he looked at me. It was a look so piercing, so full of grief, a look so *human*. I almost laughed (I felt too sad to cry) to think that there are people who do not know that animals suffer. But most disturbing of all, in Blue's large brown eyes was a new look, more painful than the look of despair: the look of disgust with human beings, with life; the look of hatred. (Walker 1989; 4–8; Kessler 2019: 99f.)

In his extensive analysis of Walker's account, the philosopher Neil Kessler seeks to understand the reasons why "what might have been a developing close relationship between Walker and Blue has stopped", noting that "the failure of close relations was not because Blue didn't have the qualities and capacities to support closeness". In particular, Kessler takes issue with Walker's attribution of the failure of closeness to "Blue's grief turned to hatred and disgust for all of humanity" (2019: 103f.).

Kessler's interpretation is worth quoting, not least because of what he has to say about responsibility, a concept that, for lack of concrete examples, tends to remain abstract when philosophers discuss relations between human and animals. He writes:

> I think closeness failed to develop in this case in part because Walker *failed to act on Blue's behalf* as an individual human in relationship with him, the individual horse. By generalizing the object of Blue's disgust and hatred to all humanity, she abstracts and thus masks its very real, tangible and personal targets. Such strong negative feelings are never, at least at first, abstract or general. All three humans – Blue's owner, Brown's owner, and *Walker herself* – fail to act substantively on Blue's behalf, to ease the suffering that was so obvious, and so obviously a result of their actions or their lack. Blue hated those three people and they failed to relationally respond to that hatred in any ameliorative way. This is precisely what I meant when I suggested that human-nature relationships are flawed first at an intimate or personal level. Rainforest deforestation occurs because that very same suffering is ignored on an individual-by-individual basis until it reaches some massive, biome-disrupting scale. Just the same, clearcutting in national forests continues because land managers charged with the care of those places fail to recognize the possibility of these individual-level close relationships and the suffering caused by their fracture or prevention.
>
> (Kessler 2019: 104)

What Kessler is getting at is Walker's responsibility to Blue as a relational partner. That *her* responsibility for *him*, for taking action to support him when he so clearly needs her to help him, has the status of a *fact* for both of them is something Blue actively contributes to when he – full of pride and happiness – shows Walker how much he trusts her and cares for her taking part in his joy by waiting for her and then bringing his newly acquired mate to the fence to meet her. Blue's decision to do that – to invite her into what matters most to him – attests to the close relationship between the two of them as built in their shared past. Yet it also carries an expectation about that relationship's future, namely that Walker will do whatever it takes to show herself worthy of his feelings for her and the trust he puts in her in that his falling in love with Brown and fathering a foal with her is a life-changing experience to him that, in addition to the pride on behalf of his now brilliantly displayed "inalienable horseness", also carries a distinct vulnerability: he might lose the love and happiness his finding a partner has given him. Dyadic relationships deserving the name "close" involve the expectation that should one party need help, the other will intervene. Why shouldn't this apply to the relationship between a human person and a horse?

As it turned out, Blue *did* lose the love and happiness that he experienced. He lost it for no reason to do with his actions, or with Brown's. Rather, he lost his new relationship by way of it *being taken away from him*, by some particular human individuals who decided to do so, and who in doing so showed no regard for his (or his partner's) feelings, no regard for what separating them meant for them. Brown had "been put with him" so that they could mate and she conceive. Once accomplished, they were separated, never to meet again.

Kessler faults Walker for "doing nothing" when this separation happened, "save a tepid inquiry into Brown's fate" that brought no result, letting the matter rest there (ibid.). Her interactions with Blue as a relational partner, he says, "become a *using* of Blue and his grief that may be to a much less abusive degree than the using of Blue by his or Brown's owner – or the using of a forest by land managers – but it seems to me a difference of degree, not kind" (ibid.). Indeed, the potential for a close human-animal relationship did exist between Walker and Blue, and the deeper lesson here, Kessler suggests, is that "it is her (and all of our) failure as modern humans to hold up our end in such encounters – to see the more-than-human self as a Self – that so often derails such closeness from coming to fruition" (ibid.). Specifically, that Walker does not "act in a way that would benefit Blue – and that Blue would be able to identify her as having undertaken – is the real failure" (ibid.). Kessler hastens to add that he doesn't mean this as a criticism of Walker as a person; his point is "to show how deeply embedded in all of our thoughts as modern human beings is that we cannot act" (ibid.).

We will never know if Walker could have brought Blue's mate back. Whether she agrees to it or not, Walker is part of a modern world "constructed to treat more-than-human beings such as horses as slaves", as she herself as well as Kessler points out. But, insists Kessler, Walker could have given Blue's owner an earful anyway, opening up an alternative scenario where:

> Blue would have seen her yelling at the owner and pointing at him. He might've *felt* her care. And perhaps in response to witnessing this, instead of Blue arriving at the fence by the apple tree the next time with hatred and disgust for a humanity that included her, he might've come there to rub that great viola-case of a grief-stricken head against her shoulder to salve his broken heart with one who cared enough for him to protest on his behalf – to feel *with* him instead of *about* him. In interdependence theory, all the thoughts and feelings in the world won't amount to a hill of beans without the third vector of *action*.
>
> (Kessler 2019: 105)

According to Kessler, then, it is an undisputable fact about the relationship between them that Walker is responsible for helping Blue; and it is a fact that she lets him down. Note that for Kessler, responsibility is an objective (factual), yet subjectively felt and confirmed, feature of that human-animal dyadic relationship, just as he would surely – were he to consider the comparison – say that it is when the parties are made up of two humans, say, Walker and her husband, or Walker and a child she comes to know. If anything, responsibility weighs comparatively heavier on Walker when her "other" is the horse Blue. Why? For the Levinasian reason that the horse is vulnerable and dependent on her in a manner that her fellow human other is not, though a difference in degree should be acknowledged between the adult other and the child other. Blue, that is, is at the mercy of what humans decide to do with him – to him – to such a degree as to make for a strong asymmetry in his relationship to *all* humans, be it Walker, his close friend, or his owner, who doesn't appear to have entered into an I-Thou relationship with him at all, but instead an I-It one, at least when judged in light of the event referred to. As far as responsibility is concerned, then, the asymmetry – essentially: in power to influence what comes to pass with Blue – in no way casts doubt on the *reality* of the responsibility, or detracts from it or makes it conditional; to the contrary, the more borne out by the facts the asymmetry is, the stronger the call – or burden – of responsibility weighs on the agent.

There is a second aspect of Kessler's analysis worth dwelling upon, even more fundamental from a philosophical point of view. This is the way he assumes that Blue would have responded had Walker decided to "yell at the owner" and to point to him in doing so. He would have felt her care. Here, Kessler is assuming a sort of transparency in the two-way communication between Walker and Blue; or more precisely, and complexly, he assumes that in the triad involved in the scene, Blue would be perfectly able to interpret what Walker communicates to his (nameless) owner as – simultaneously – a communication *from her to him*. More still, Blue is portrayed as getting it perfectly right about what that communication is *about* – i.e., that Walker's intention in yelling at the owner is to criticize what he had done to Blue in such a way as to demonstrate her support for Blue, her endeavour to show herself – to all concerned – as his friend and as enacting the responsibility that goes with such friendship – this as a matter of course, at that, as the stuff of which such friendship is made, as will be apparent to all. This being so, there is nothing "mysterious" about this sort of communication being exchanged between a human person and a horse (recall the case against scepticism offered by Midgley and by Crist in Chapter 1). This of course also holds for the scenario that is the all-too-real one in this case, where Walker fails to intervene on Blue's behalf. That failure Kessler also sees as communicated from Walker to Blue, and as understood by Blue for what it implies, causing him to show her his "hatred", only she chooses to interpret it as "the look of disgust with human beings, with life", thus deflecting the fact (following Kessler) that Blue's hatred targets *her*, not the abstract generality "human beings" she takes refuge behind, intellectualizing the experience and denying her agential role in it. The question she thereby refuses to face is whether Blue will ever be able to forgive her.

Kessler's overarching concern is what he theorizes as "the failure to acknowledge the thoughts, feelings and subsequent actions of more-than-human-beings", a failure on the part of humans that he, like Shepard, maintains is a historically recent development. Since Kessler's take on this is more philosophically rigorous than Shepard's, we shall take a closer look at it.

What do we mean when we say that "many animals arouse our sympathy" or that "landscapes full of variety appear to us spontaneously as beautiful"? The allegedly "scientific" answer, says Kessler, is that features (qualities) such as beauty are seen, and causally explained, as emerging unilaterally from the human, instead of as a product of the human-nature relationship such that what happens is that we respond to the thoughts, feelings, and actions of the more-than-human beings *present in them* and communicated *by* them *to* us. As humans we need, and indeed have, what Kessler calls "living Selves", those of our various "others"; living Selves, not "lively matter", as he says, castigating posthumanism and new materialism for the "dearth of individuation", i.e., their failure to acknowledge ontologically and conceptually "individuals as individuals-in-relation" (2019: 137).

On this basis, Kessler criticizes the error of seeing culture as a product of human contrivance – that it is purely *social* – instead of human-nature relational as well. What we risk leaving in place, unexamined, is the "ontological stance that humans and more-than-human beings are foundationally separate from each other, with human culture fully forming prior to being influenced relationally or otherwise by the more-than-human world". Kessler's contention, then, is that "humans don't see the more-than-human world through a lens of socially constructed culture, they see it through a lens of human-nature relationships first, despite all attempts to conceptually eradicate the latter's influence and existence in modernist views" (2019: 274).

It is tempting to consider what Kessler is saying as simply repeating earlier points. But that would be a mistake. What Kessler explores is not the notion of anthropomorphism that we discussed in Chapter 1, but the *link between ontology and culture* that is required to explain the power that anthropomorphism continues to exert, notwithstanding the many dismissals of its dualistic origins in Cartesianism. According to Kessler, we are still, as a culture and as a society, in a situation where "the more-than-human beings with whom we are in relations must in actuality be a kind of relational *tabula rasa*, forming only the backdrop onto which humans, by drawing from some isolated locale inside themselves, project their concepts of what more-than-human beings are like. By extension, the more-than-human beings themselves have no-more-than material nature of their own to be perceived" (2019: 277); "to be perceived" meaning: for us to respond to, as the qualities – feelings, thoughts, actions – that we experience *in*, say, animals such as the horse Blue. Kessler points to the dualistic, *post hoc* conceptual movement away from close relational interchange with more-than-human beings. As a result of this operation, everything that emanates from both animate and "inanimate" more-than-human beings and our close relations with them "is largely lost to modernity". This loss has happened gradually, not suddenly and all at once, so that authors like Wordsworth and Thoreau become a weigh station along this path to the present, descriptive of a time when "modern men still *felt* more immediately unpleasant [or pleasant] things that more-than-human Selves were feeling and conveying to them" (2019: 278).

There is a bottom-up as opposed to top-down logic to Kessler's argument that may seem simplistic, even naïve given the broader context where systemic forces limit what a single individual can do. Granted, Walker could have done more to protect Blue from the misery of being treated as a mere means by his owner. But the practice where the match between Blue and Brown – horses like them, everywhere – took place, facilitated for purely human ends, not the involved animals' ends, would continue unabated, even if some kind of arrangement – by way of exception – were made for Blue following an intervention from Walker. Recall Kessler's statement that "Rainforest deforestation occurs because that very same suffering [as Blue's] is ignored on an individual-by-individual basis until it reaches some massive, biome-disrupting scale" (2019: 104). Does clearcutting continue because land managers "fail to recognize the possibility of individual-level close relationships" on the ground (ibid.)?

The causality suggested by saying that clearcutting continues "because" individual failure is certainly simplistic. But Kessler's approach is that of an ontologically oriented philosopher, not that of a sociologist, economist or political scientist; the systemic power structures and imperatives of growth and profit endemic to present-day global capitalism are not his topic. Nonetheless, he has a point worth making: the level on which an individual person's encounter with a particular animal would make a difference cannot, and should not, be ignored – where ignoring it takes the form of abstracting from it, suppressing it or – simply – preventing it from occurring. The *modus operandi* of the Nazi genocide of the Jews was a product of a deliberate attempt to replace closeness between perpetrators and victims by distance: no longer seeing, hearing, smelling, and touching those to be killed, it becomes easier to take part in doing so (Bauman 1989; Hilberg 1985; Vetlesen 2005). The human psychology according to which one-to-one (first-hand, physical, multisensuous) encounters take place and give rise to hesitancy and qualms over killing – *me* killing *that* other person – is principally the same where the other is an individual animal and not another human person: the sheer experience of encounter makes for a bond, involves a commitment, forces the agent (me, not somebody else) to decide what to do: to heed the appeal to be protected issued by that other, or to disregard it, pretending it isn't issued or that it has no authority. Perhaps it boils down to this: would the practice proceed if the individuals participating in it – the mass killings of Jews, the deforestation of the rainforest (and I do not mean to put the two on a par, morally speaking) – allowed themselves to experience those human and/or animal others affected by their actions as so many distinct *individual* humans and/or animals?

Kessler does not raise the question quite so dramatically. Instead he cites the seemingly less fatal practice of bird banding, often justified by reference to how it helps the birds materially survive. He is quick to point out the self-serving logic at work here, relying on denial and bad faith: if the aim were to arrange things for the benefit of the birds, to relieve them entirely, there would be no need to determine via banding just which pressure is most disadvantageous to the birds in the first place. The scheme as such is circular as far as cause and effect go. What it helps put on display is "what we modern humans find acceptable to do to the birds inside of a largely dualistic and anthropocentric frame of reference. By carrying on in our relationships with *this* mist net and *that* bird caught in it without a thought as to how *that bird feels* to be treated thusly by us, is to simply continue imposing the same old dualistic human-nature relationships" (2019: 280f.). In fact, it's quite straightforward: "We cannot have our highway system, our smart phones, and even our wind power and let birds live how they will live. Not how we *think* they need to *materially* live – but how they will *choose* to live *for themselves*" (2019: 280).

5 Responsibility and grievability in the face of extinction

I

> A jaguar ventures out of the cover of trees. She, too, is under the close eye of human researchers, tagged and tracked, as part of an attempt to halt the fast downward trend of jaguar populations across the Americas. Tonight, she is hungry and is seeking unusual prey – sea turtles. Until recently, she had no need to hunt this well-protected quarry. The forest was large and held many options for her. She could take peccaries, monkeys, agouti, or many different kinds of birds or fish. But the forest grows smaller, and so do her choices. […] Learning to kill a turtle also involves learning its temporalities and spatialities, being in the right place at the right time (synchronicity), hoping for prey that is both available and reliably so. How many turtles will survive and return next year? How much habitat will the jaguar have left? And what are the conservationists to do when one endangered species starts eating another?
>
> (Bastian 2017: 160f.)

There you have it, in a nutshell. Michelle Bastian's report from Tortuguero beach, on Costa Rica's Atlantic coast, brings the various elements together: wild animals becoming fewer, their traditional prey being replaced by new ones, with humans monitoring the predator's every move, wondering what the future holds in store, knowing – as do the animals too, or don't you think so? – that "the forests, the peccaries, the jaguars, the leatherbacks – all are under threat" (ibid.).

The threat referred to is that of extinction. Extinction is not a novel phenomenon in the history of the species. What is novel, indeed unprecedented, are two things: the sheer pace at which extinction of animals and the shrinking-cum-disappearance of their natural habitats take place, and the role played by humans and our activities in causing the extinction.

The current form of extinction has given rise to a new interdisciplinary field: extinction studies. Often individual researchers will specialize in one particular threatened or critically endangered species; Bastian's is leatherbacks. In the same vein, comprehensive studies will typically have a chapter

DOI: 10.4324/9781003317692-6

for each animal species considered. Irus Braverman's *Wild Life* is a case in point, with a chapter dedicated to each of the following animals: the Puerto Rican Crested Toad, the Sumatra Rhino, Golden Lion Tamarin, Black-footed Ferret, Puerto Rican Parrot, Red Wolf, and Tasmanian Devil.

I mentioned in Chapter 4 that animal rights theorists focus on individual animals, as distinct from conservationists whose focus is species-oriented. Historically, this has been regarded as a clear-cut distinction, and notwithstanding their philosophical differences, the two approaches have been able to coexist in a peaceful manner, with no need to declare the one more valid or more urgent than the other.

This situation is currently changing, however, as a result of the pressures on the various animals and their habitats that now threaten an increasing number of them in their natural habitats. Questions never asked before are now being raised with a sense of urgency nobody has been prepared for, be it conceptually, ethically or in terms of answering the series of "what to do" that knock on the door. Of course, many of these are questions about priorities, about what sort of animal to regard as more important to protect than others, ultimately to save from going extinct. But what does "important" mean here? Important in what respect, in virtue of what, and for whom?

Recall the formula I have used on several occasions: that the constellation we have to pay attention to, is that of the *ecologically required* and the *morally right*. The question is: what if the two are in conflict? Can it ever be morally right to sacrifice individuals for the sake of a species, given that doing so is ecologically required in the sense of being necessary to protect the habitat or the ecosystem?

Before I engage with these questions, it is useful to say more about the context in which they nowadays arise. As indicated, I will concentrate on the situation of endangered species of animals.

Traditionally, conservation efforts can be undertaken either *in situ* or *ex situ*: either the animals are sought maintained and protected in their habitat in the wild, or they are taken into captivity and sustained there, typically in a zoo. This either/or distinction has been regarded as clear-cut, the difference between the one setting and the other as evident, and in the typical case, conservation has taken place where the animals in question exist in the wild, without removing them to another geographical place, let alone into captivity.

As Braverman explains, this is changing. Conservation is currently morphing into a continuum between the two poles. Endless combinations of *in situ* and *ex situ* spring up almost daily, leading Braverman to describe no less than seven *inter situ* nodes: gene banks, zoo breeding centres, conservation farms, conservation hatcheries, protected areas, wildlife refuges, and national parks (2015: 17f.). As fluidity and hybridity come to replace the old quasi-physical division between *in situ* and *ex situ*, the dichotomy underlying the terms loses the plausibility granted it for so long, that of being true to the facts, such that *in situ* would correspond to wilderness and hence to nature, and *ex situ* to captivity and hence to culture. This being so, the conventional division of labour with field biologists and wildlife managers on the one hand, and zoo and lab experts on the other is also being challenged; the involved professions and disciplines now need to collaborate in a more integral way than ever before if their shared goal of conservation is to succeed. Despite these changes, Braverman maintains that the old dichotomy still makes sense: "Without a wild, free, and pristine nature, conservation in captivity is meaningless; and without the notion of captivity, wilderness as the very opposite of captivity cannot exist" (2015: 4). Nevertheless, we need to acknowledge that "there is not one nature but many", as part of the attempt to understand why the existing habitats of a growing number of species are becoming less viable. In reality, then, what conservationists refer to as a natural habitat must often be actively managed alongside the construction of an alternative one, making for the continuum indicated above. With increasing frequency, "the last handfuls of animals from a species on the brink of extinction are transferred into intensively managed sites to ensure their survival"; in this way, "a livable nature is constructed in captivity for the conservation of threatened life" (2015: 10).

Probably as much by inclination as by training, conservationists have been committed to a hierarchy where *ex situ* conservation is seen as a third-best solution, after the preferred no-management-at-all approach and the second-best *in situ* management that occurs, say, in big national parks. Misgivings about *ex situ* conservation, be it in zoos or in labs, still persist, based on the notion that *animals cannot behave naturally in captive settings*: in trying – having – to adjust to what may objectively be described as an "artificial setting", the animals lose their distinctive naturalness, depending as it does on their living their entire lives in their natural habitat, which traditionally means in the wild, undisturbed (largely if not entirely) by human intervention and facilitation. Accordingly, captivity cannot but produce "artificial" animals, prompting questions like: (1) If captivity is the only option left to save the animals in question, is it thereby right?(2) Is there a sense in which these particular animals are lost the moment the last individual is moved from its natural habitat and into captivity – to be sure, its survival may be ensured that way, but is the animal that survives in an artificial setting the same animal that used to live in a natural one?

Indeed, how to manage species that are extinct in the wild and now exist only in virtue of efforts made to allow them to exist in captivity? As Braverman explains, many conservationists hold that select species that would otherwise go extinct are to be maintained in captivity "until their survival can be secured in the wild", so that conservation in captivity is justified "only when conservation in the wild is no longer a valid option – namely, as a measure of last resort" (2015: 71). Interestingly, Braverman observes that it is becoming more common than before to hold that captive populations are legitimate and even necessary "in and of themselves, and not only for supplementing the wild": since surviving in the wild is becoming increasingly precarious for an increasing number of animals, captivity would benefit the animals in question by "saving them from the stress of surviving in the wild" (ibid.).

But doesn't such reasoning involve a slippery slope? Until now, the dominant view has been that the very point about wild animals, the value peculiar to them, is that they should be left to their own devices, their being wild being tantamount to their *being free in the wild*, undisturbed and unaffected by any sort of human interference. Of course, this throws up the issue of whether there is, in the era of the Anthropocene, any such thing as "wilderness" in the pristine sense traditionally implied. And if that quality is gone, due to human activities that alter the very make-up of the natural world, disrupting its ecosystems and their regenerative capacities, then many people would say that the very distinction between wild animals on the one hand and domesticized animals on the other has to be abandoned: even with respect to what remains, say, of iconic predators like lions, tigers, jaguars, and leopards in a setting where they are not fenced in and thus "held captive", theirs will now be an existence thoroughly shaped by human activities and decision-making – not necessarily in the form of humans directly (physically) interfering with them in their day-to-day movements but in the sense that their staying in the (quasi-) wild – often in national parks – is now a matter of all-around-the-clock human monitoring of the animals and their every move, so that the lives of the animals are drawn into the use of ever more advanced technologies (tagging them, filming them from helicopter, injecting them so as to move them from one place to another), amounting to an enormously resource-intensive undertaking on the part of humans for the sake of "saving the last wild animals qua wild", whereby the means employed to reach that stated aim cannot but spoil it.

The end point of this trend is that the destiny of animals in the wild is wholly at the mercy of humans: since we are fully capable of ruining (what's left of) the natural environment that their being wild relies upon, only a decision on our part to *abstain* from that destruction will ensure, for the time being, their survival. For all the drama of this big picture, the slippery slope I alluded to kicks in with varying degrees of drama in each separate case: the more stressful the non-captive form of life may appear to become to certain animals, for regrettable human-made causes at that, the more tempting it may become to decide to move them into captivity in some form, tempting because seemingly the most "realistic" and so "responsible" thing to do, everything considered. Captivity will then be chosen over existence in the wild, for the sake of saving the last "wild" animals. Subtly or not so subtly, then, captivity has become the default stance, the new normal for the animals in question. For some, this will be accepted as the "last resort" Braverman talks about. For others, it will be rejected as a contradictory, if not outright self-defeating scheme, a perversion of the very mission of wildlife conservation, a case of giving into, and becoming complicit in, the very forces that must be resisted.

Although there is often disagreement about *when* the situation has become so dire as to justify the transfer of individual animals from *in situ* to *ex situ* settings, there is an increasing agreement among conservation experts that captivity can and should serve as the last resort for a species on the brink of extinction. *Life in captivity is better than no life at all*: that is the bottom line. There comes a time when discussions about whether enough was done to prevent the transfer into captivity from becoming the "only realistic and responsible thing to do" become moot, of academic interest only. If this is the current consensus among conservationists, it is a position strongly opposed by animal rights theorists and activists alike. They question what they take to be the privileging of the interests of species or populations over what they maintain are the rights of individual animals, rights that should be trump precisely as individual such pertaining to each particular animal.

Tom Regan states the argument like this: "The general policy regarding wilderness would be precisely what the preservationists want – namely, let it be! Were we to show proper respect for the rights of the animals who make up the biotic community, would not the community be preserved? And is not that what the more holistic, systems-minded environmentalists want?" Rights, Regan submits, pertain to individuals only; "species are not individuals, and the rights view does not recognize the moral rights of species to anything, including survival" (Braverman 2015: 77; Regan 1983: 363, 359). Sparking much controversy, Regan has gone so far as to describe the requirement that individual organisms be sacrificed for the whole as "environmental fascism" (ibid.), citing Aldo Leopold's "land ethic" as a case in point and charging that holistic approaches as such are stuck with the danger of fascism.

I shall come back to the philosophical implications of Regan's position below. For the people engaged in the decision-making about what to do in the particular case, the way an animal rights theorist like Regan plays the individual and the species out against each other is not helpful. The case of the California condor may serve as an example. In the late 1970s, there were 22 birds remaining in the wild. A decision was made to capture them all; they were all subsequently put into the captive propagation programme. The assessment of the federal agency in charge was that "if we had left the condors in the wild, they would have gone extinct". While the birds were being bred in captivity, the ecological context was, to some extent, regained for them, so that when they were released into the wild they were able to expand their range and number. Indeed, following the reintroduction of the captive California condor into the wild, more than 300 condors now inhabit large areas of Arizona and California (see Braverman 2015: 77).

This is no doubt a success story, made possible by a sensible and informed pragmatism on the part of the personnel involved rather than by the attempt to settle a philosophical dispute. Even so, those engaged in this sort of decision-making will ask themselves tough questions about what they are doing given the bigger picture, such as: "Is it better to vanish in the wild than to lead a rich life in captivity?" "Would it be more desirable to die free rather than to live in captivity?" (2015: 80). Can one say that "the condor was destroyed in order to save it" (Marris 2021: 112)?

Note that these questions are raised in a manner referring to individual animals, inviting us to assess what alternative options would look like from their individual point of view, as it were. Perhaps this "methodological individualism" is inevitable, besides being taken for granted, representing the culturally and institutionally entrenched individualism – conceptual and ontological, moral and legal – that I have pointed to on several occasions. And to be sure, questions concerning the "desirability" of one alternative over another – say, captivity over wildness – do seem to make sense to us only as applying to *particular individuals*, rather than to a species: not only because a species consists of so many individuals whose points of view may diverge, presumably the more so the more advanced and rich in capacities for experiences the individuals making it up; but also because something – a kind of life, in a certain kind of setting – can only *be* desirable to an individual creature, not to a species qua species.

It does seem, then, that a strong argument can be made that *the locus of experience*, of having the difference between pleasure and pain, flourishing or suffering, matter, is the individual not the species, the particular member not the (biotic) community. If we grant this (and it does seem compelling), does it follow that the individual is always trump, always to be given primacy? Primacy over what? Well, the species, understood as the whole, as the entity that, were we to give it primacy, would thereby crush the individual since leading to the sacrificial logic that we saw Regan warn against?

There is something artificial and overly cerebral about this way of pitting individual and species against each other, as were the essential relationship between them that of conflict, of being at loggerheads, and concluding that the choice between granting trump status to the one rather than to the other equals the choice between right and wrong. Are the alternatives that clear-cut, simplistic, and mutually exclusive on the ground? Is this theory at all suited for the practical decision-making and tough choices conservationists are presently confronted with?

To answer these questions, we need first to note that on Regan's theory, not all animals have moral rights. That not all animals are to be considered as morally equal is a view Regan shares with Singer, whose plea for equal consideration comprises the diverse interests of all *sentient* animals. Regan for his part is even more restrictive: only those animals who qualify as possessors of "inherent value" should be granted rights – that is to say, only animals who satisfy the "subject-of-a-life" criterion. To be such a subject requires being self-conscious and having the capacity to believe, desire, conceive the future, entertain goals, and act deliberately. This is certainly a tall order. Truth be told, most animals have no chance of qualifying. The class Regan is left with is an exclusive elite among animals, namely "mentally normal mammals of a year or more". As critics such as J. Baird Callicott rightly point out, the exclusiveness of the category of animals for which Regan, given his own definition, actually makes his case is masked by his book's title: it should read *The Case for Mammal Rights* (Callicott 1989: 40).

As a matter of fact, the vast majority of endangered species are not made up of the rights-holding mammals prioritized in Regan's theory; the vast majority are plants and invertebrates. Factually, then, his is a theory of animal rights that has nothing to offer to most of the creatures making up other-than-human life on earth. Conceptually, Regan's theory is weakened by buying into the symmetric model I have critiqued above inasmuch as his requirements for qualifying as a moral addressee (in his case: a bearer of rights), highlighting as we saw overly cognitivistic capacities, not only preclude most animals; more fundamentally still, in my view the requirements are such as to be necessary and relevant to being a *moral agent*, not a moral addressee. Presumably, the whole point about making a case for the importance of introducing rights into the domain of animals and not only humans (Regan's very project) is to advance their entitlement to *protection*, that is, to ensure their status as beings that should not be treated as mere means, as beings available for abuse, exploitation, and, ultimately, extinction by those who treat their lack of consciousness- and cognition-based agency as a licence to abuse them with impunity. It is surprising that a critic like Callicott does not pay any attention to this point about how Regan conceives the relationship between agent and addressee.

Whereas Regan and the generation of environmental philosophers he belongs to, alongside Callicott, Paul Taylor, and Holmes Rolston, primarily address animals in the wild, the "logic of sacrifice" that conservationists and ecophilosophers alike are concerned with today is increasingly focused on the issues raised by captivity. Thom van Dooren, one of the leading scholars in "extinction studies", puts the relevant difference like this: "While it should not be forgotten that free-living birds also experience a range of deprivations and stresses – the world beyond the fence is not Eden – this does not change the fact that life as a captive breeder remains a kind of 'sacrificial life', a life given, and not by one's own choice, for the good of others. It is a life sacrificed as part of an effort to care for a species" (2014: 114). Working on a project to help ensure a future for the endangered Whooping Cranes, van Dooren reminds us that it is not only individual Cranes whose lives are drawn into the fray: "In a range of ways, a host of individuals of other species are also positioned as sacrificial surrogates for the continuity of the Whooping Crane" (ibid.). To illustrate, in order to do the tests necessary to establishing the feasibility and best practices for ultralight aircraft-led assisted migration, major hazards included predation by Golden Eagles and collision with power lines and the propellers of ultralight aircraft; many birds, of different sorts, were injured or killed in these tests. In this situation, experimenters will sometimes deliberately select "hazardous" migration routes with hundreds of power-line crossings to thoroughly explore the nature of the potential danger and "trouble shoot solutions before beginning with the endangered cranes" (2014: 115).

I have not seen the word "cannon fodder" used in this literature, probably rightly so. But it alerts us to the ethical dilemmas involved: Can some animals be used as a means to further the ends of others? What are the criteria suited to the prioritizing involved? That the "means" animals are less important than the "ends" animals? What would justify such ranking? In light of all the energy spent to reject considering humans as "higher-standing" than animals of all sorts, condemning it as self-serving, dangerous, and blatantly unethical, do we not risk repeating this selfsame logic of hierarchy were we to engage in schemes deeming some animals so much "higher" in importance and worth as to justify putting the lives of so many lower-ranking others at risk? Or, finally, is it more about numbers than qualities, such that we may lower the ethical bar whenever the "means" animals are (comparatively) plentiful and the "ends" animals so few as to be critically endangered qua species?

Conservation efforts seem to have moved along a continuum scarcely anticipated by the generation of philosophers who made their first forays into the field of animal rights, animal ethics in the 1970s. At that time, the primary task was pretty straightforward: to argue the case for the establishment of national parks big enough to allow the animals to persist in their original way of life in their natural habitat, left to their own devices qua wild, where the preservation of that wildness – conditional on the (relative not absolute) absence of human interference – is the quality to be protected, valuable in its own right, such that the point about these animals is that *they are not put to any sort of use by humans*, and for human-directed purposes.

The continuum I alluded to is one where, gradually, bit by bit, place by place, species by species, the original way of life of the animals in their natural habitat as such is becoming an increasingly rare and so endangered entity, a situation calling for measures and methods never before entertained because simply not required. Accordingly, those animals still left in what remains of the wild are less and less left to their own devices as they are being more and more monitored, with ever-expansive advanced technologies making sure that, say, not a single wolf in my native Norway is left unseen and unmarked for as much as a single day of its life cycle, be it in the so-called "wild". Then, there is the transfer to a zoo and so captivity of animals so few and far between in the fragmented, degraded, and vanishing wild as to be considered "unsustainable" in their original setting, whereby captivity becomes the only realistic option for what used to be iconic wild animals, such as the rhino, and perhaps tomorrow the orangutan. This being so, is it not tempting to consider the employment of other animals, less threatened and richer in number of individuals, to assist in efforts to save those in serious trouble – even if giving such schemes the green light involves the calculated risk that a considerable number of them are killed in the process? When is it worth it, and when is it not?

Indeed, can a "sacrificial logic" where individuals of "least concern" are treated as available and expendable forms of life in the service of other, ostensibly more important and valuable beings, ever be justified? Who is to be considered as "killable" for the sake of the greater good of conservation? And if it strikes us that we would instinctively find this kind of reasoning utterly problematic were human individuals the creatures involved, how can we possibly find it acceptable in the case of animals?

Drawing on Donna Haraway's reading of Derrida, van Dooren argues that "refusing this sacrificial logic is about insisting that these individuals [deemed as killable] *do* make ethical claims on us. The point is not that we can never kill or cause suffering in the name of conservation, but that the decision to do so...should never leave us comfortable and satisfied" (2014: 117). Is it worth it, then? On van Dooren's view, "even if the death of the last Whooping Crane is delayed for several more decades, even if the species recuperates up to the point that it can again be self-sustaining in multiple free-living migratory populations, the violence and suffering that has produced this outcome will not and cannot be erased or 'justified away'" (ibid.). Van Dooren will not be comfortable as long as some forms of flourishing are sacrificed for the sake of others. This, he says, is an "all-too-common occurrence in conservation projects where care for the species is usually uncritically taken to trump all other ethical concerns". The closest he comes to arriving at a conclusion is his statement that "refusing a sacrificial logic requires us to develop ways of caring for cranes that take seriously the flourishing of both species and individuals", and so exposing us to "an imperative to constantly strive to do better than we are now" (2014: 118). As Haraway says, we need to stay with the trouble and not pretend there are answers that come without dilemmas, solutions without casualties. The ecologist Chris Thomas, commenting on a project in New Zealand involving the poisoning of a huge number of rats so as to prevent the demise of kiwi, articulates an important truth when he says that "the rat doesn't want to die of poisoning; the kiwi doesn't want to be attacked. I don't think there can be an ethically correct answer. We can trade off the deaths of one thing versus the deaths or survival of other things. But there can't be an ultimate right or wrong" (Marris 2021: 199f.). Decision-making in this area has to be carried out on a case-to-case basis.

It is noteworthy that van Dooren, a philosopher by training, completes his discussion of this matter without once referring to the notion of rights. Would an ethics that propagates animals as bearers of rights enrich his discussion, not to mention help him arrive at a more rigorous conclusion?

I don't think so. In saying that, I have Regan's version as well as its logical alternative within rights theory in mind. Above, drawing on Callicott's criticism, I faulted Regan for prioritizing an "elite" class of animals (i.e., cognitively sophisticated mammals) over all others, in effect securing – at least in theory – respect for their "inherent value" for what in reality is a small minority of all animals, leaving the majority by the wayside since falling short of qualifying for moral status. The ranking of higher and lower within the universe of animals is in danger of creating its own kind of apartheid, at least as long as lack of inherent value is tantamount to availability as mere means to others. Thus the binary schema of ends-in-themselves vs means-for-others that animal rights theorists ever since Singer's ground-breaking *Animal Liberation* have been intent on attacking as a crucial element in their critique of anthropocentrism, often taking Kant's version as their primary target, is not abandoned, let alone overcome, but replicated, only now with regard to the relationship between (different species of) animals instead of that between humans and (all) animals. Whereas this poses a challenge for Regan, as far as Singer is concerned, and as commentators have pointed out ever since his breakthrough as the leading animal rights theorist, he may be charged with perpetuating a dualistic outlook where living beings are either included in or excluded from the all-decisive, one-criterion-only category of sentience. Accordingly, his position does not question the notion of the merely instrumental value of nature *per se*, the very notion that has been the call to arms for generations of ecophilosophers. Far from taking part in that battle, Singer's theory is content to merely move the borders of the category held to include those who qualify for moral consideration, i.e., as addressees in my terminology.

What about the logical alternative – namely, to posit that all animals are equal as far as their inherent value is concerned, so that no species-based discrimination between them can be justified, and an end can be put to practices involving sacrifice? If all sorts of animals are morally equal in the sense of inviolability, so that it will never be morally justified to treat somebody as mere means, and to determine that the interests of some other(s) be given primacy, then where does this take us in terms of decision-making? If every step on the grass is morally problematic in that it harms some of the organisms found there (to allude to a popular reading of Albert Schweitzer's "reverence for the sanctity of all life" doctrine), then how can we move – make decisions – at all? If the problem with Regan's position is that it allows for the treatment as means of animals that we intuitively feel should not be that readily sacrificed for the sake of "higher-standing" others, the problem with Schweitzer's is that it forbids our every step, every move, inasmuch as there is no such thing as moving around – be it humans or animals that do so – without some instantiations of life, some loci of conatus, being harmed.

I conclude that granting animals (or some of them) rights is not the ethical position best suited to tackle the issues at hand. To be able to explore a more promising approach, it is necessary to look more closely at the relationship between individual and species.

II

Above I criticized Tom Regan's depiction of the relationship between individual and species as one of conflict, imposing upon us the choice between prioritizing either the one or the other. I now submit that to ask: What is trump, individual or species? is to start the discussion from the wrong place, ethically as well as ecologically.

We need to gain a better understanding of what an individual and a species is, one that recognizes their interconnectedness, the fact that the one cannot be, cannot persist, without the other. Van Dooren's approach is more fruitful than Regan's, since more ecologically attuned: "Each species lineage embodies a particular *way of life*, a particular set of morphological and behavioral characteristics that are passed between generations" (2014: 27). He hastens to add that "this is not a static way of life. More than the sum of those individuals currently living, species are engaged in an ongoing intergenerational process of *becoming* – of adaptation and transformation – in which individual organisms are not so much 'members' of a class or a kind, but 'participants' in an ongoing and evolving way of life" (ibid.). In this context, "any individual bird is a single knot in an emergent lineage: a vital point of connection between generations – generations that do not just happen, but *must be achieved*" (ibid.).

In this way, then, the species Whooping Cranes needs the participation, the contribution made, by each and every individual Whooping Crane: as so many parts of the whole they constitute, so many members keeping it afloat in time and space, helping sustain it as the peculiar species it is, no individual Crane is expendable. The *way* it proves itself as non-expendable, however – *pace* Regan – is qua means to the sustaining of the whole without which it qua individual would not itself thrive and persist: only by being an instrument in this way – being instrumental to the whole it partakes of – can it sustain the value that philosophers, one way or the other, some more cerebrally than others, wish to claim on its behalf qua individual. In other words, to be an instrument and to have value in itself are not contradictory statuses or roles; rather, they are mutually constitutive, the one being the condition of possibility for the other. There simply is no species without thriving individuals; no thriving individuals without the persistence in time and space of its species, the one trajectory being inseparable from the other – hence also endangered the moment the other is. To pit individual and species against each other is to be ecologically illiterate.

At this point, conservation biologists will typically observe that whereas individuals in plural make up a species, single individuals are still expendable without endangering the existence of their species (Sverdrup-Thygeson 2021). While this may be true in theory, it is not always true in real life – it depends on what turns out to be the current plight of the species in question. This being so, to treat, in practice, individuals per se as expendable may prove fatal: the last remaining rhinos are not expendable but *irreplaceable*, factually so, whether our philosophical position recognizes them as such qua individuals (specimen) or not: their death will be tantamount to the end of the species, closing down its historical trajectory forever and thus representing a true loss.

Van Dooren makes the overall point about interconnectivity in a perspective that alerts us to the importance of continuity in time and place, highlighting how it comes to be that *a species is the achievement of its individuals* and thus something eminently dynamic, something that is always changing and thus never "closes" as far as its potential qua species is concerned. Here is his description:

> Individual participants in a species may not be aware of where they have come from or where they are going in this broader evolutionary sense, and yet, it is *their* striving for continuity that achieves it nonetheless, This striving is, of course, present in the individual organism's efforts to maintain itself, but what is equally important is the efforts of living things to pass on that life to subsequent generations. Each year, taking on the dangerous and demanding obstacles of reproduction, albatrosses, like so many of Earth's other creatures, invest their lives in a generational continuity that exceeds them. [...] Bound together by shared inheritance and a shared fate, a species is always more than its current manifestation. Extinction is the loss not of a single fixed "kind", but of a potentially limitless set of emergent and branching flight ways from the present into the diversity of the future. Each species is ultimately a flight way *beyond itself*.
>
> (2014: 38)

To address this situation, which is that of mass extinction, van Dooren proposes a new ethical imperative: "to hold open space for other species" (2014: 40), that is, to ensure that other species than humanity have a future. One is reminded of Hans Jonas' version of such a new imperative: "Act so that the effects of your action are compatible with the permanence of genuine human life", or negatively put: "Do not compromise the conditions for an indefinite continuation of humanity on earth" (1984: 11).

One may wonder whether van Dooren and Jonas are on the same page. Is there a conflict between the obligation to hold open space for other species, seeing to it that their trajectories continue into the open, change-involving future van Dooren describes, and the obligation to ensure the permanence of a genuinely human life? Is the one imperative biocentric, the other anthropocentric? If so, do they contradict each other?

To raise this question is to anticipate a topic to be discussed below, namely what responsibility means in the age of extinction, by which I indicate that it cannot mean the same as before. But to return to the contrast between the two imperatives: Can we, humans in our capacity as moral agents, hold open space for other species without at the same time ensuring that we ourselves persist into that open future to be shared with all our species others? To hold that we cannot accomplish the other-directed part that van Dooren highlights without securing the inner-directed one is of course to suggest that, contrary to the conflict indicated, the two imperatives are not only compatible but depend on each other.

This reasoning may well appear convincing in theory. But the real validity test is to do with how the implied claim that we humans are needed to ensure that other-than-human species have a future on planet Earth plays out in practice. We currently know more about the answer to that question than most of us care to admit or pay serious attention to, attention of the sort that makes it imperative for us to change our ways, culturally and individually. To put it differently, Jonas may have got it plain wrong: the future of humans' many "others" does not depend on the "permanence of genuine human life"; on the contrary, what we must recognize today is that *their* future is put in jeopardy by *our* continued presence.

What do I mean by saying this? What do we do if we accept the notion that our "others", counting multiple sorts of animals among them, are better off without us?

The context for this question is succinctly set out by van Dooren:

> For this briefest of evolutionary moments, *Homo sapiens* has come into direct and profound contact with the world's albatrosses and countless other species. Our flight ways have crossed, and each of us has become implicated in the fate of the other. [...] Perhaps what is most tragic about the current situation is not the "failure" of albatrosses to adjust or adapt to new threats and an altered environment: intensive long-line fishing or brightly colored plastics that look like food. Rather, what is most tragic is another failure to adapt. Our own failure – in which some societies and some people are far more complicit than others – to come to terms with our own relatively new capacity to systematically alter environments in a way that undermines possibilities of life for other living beings and, ultimately, for ourselves. Perhaps it is we who have not yet "evolved" into the kinds of beings worthy of our own inheritances. (2014: 43)

On the one hand, owing to the role of humans (incomparably greater in the case of the richest and most consuming, most CO_2-emitting 10 per cent of the population than the rest) in putting the future of animals like the albatross in jeopardy, the answer to the question about whether they would be better off without us seems to be brutally simple: yes, they would. I am not suggesting that van Dooren is saying this – what one can say, though, is that it appears to be a logical consequence of his causal analysis, factually correct at that. On the other hand, a different response is also possible, namely that precisely because humans have been instrumental in causing the crisis affecting so many of our others, that very responsibility makes it incumbent on us to do everything we can to clean up the mess and devote our most sophisticated capacities to that other-directed task.

Even though Hans Jonas' *The Imperative of Responsibility* was written in the late 1970s, the book does, to a certain extent, anticipate the scenario that van Dooren depicts as the present-day reality. Jonas writes: "In the choice between man and nature as the struggle for existence poses it time and again, man indeed comes first, and extra-human nature, even with its own right ('Würde') acknowledged, must give way to his superior right". This, he continues, is so because "the simple rule holds that egoism of the species – each species – takes precedence anyway according to the order of life in general". In other words, "the mere fact of superior power legitimates its use" (1984: 137). Nonetheless, Jonas acknowledges that in the current circumstances we have to move beyond this prerogative of our own species. We have, that is, to "include a duty toward nature as both a condition of his [sic] own survival and an integral complement of his unstunted being". What we need to realize is that "the common destiny of man and nature, newly discovered in the common danger, makes us rediscover nature's own dignity and commands us to care for her integrity over and above the utilitarian aspect" (ibid.). To make sure to protect what has up till now been taken for granted throughout human history – "that there are men, that there is life, that there is a world for both" – the proposed new ethics, born of danger, Jonas stresses, is "necessarily an ethics of preservation and prevention, not of progress and perfection", knowing full well that man is no longer "simply the further executor but also the potential destroyer of the purposive labor of nature" (1984: 139).

To be sure, there is something outmoded and less than state-of-the-art environmental science in formulations like "the choice between man and nature as the struggle for existence poses it". Nevertheless, Jonas' attempt to reconceive the ethical imperative so that it may help orient our decision-making in the age of extinction is a promising starting point in that it recognizes what previous imperatives (say, Kant's) did not, namely that the only moral agent on Earth endowed with the capacities required for responsibility is now the most powerful one in terms of capacity for *destruction*. This coexistence of responsibility and capacity for destruction of the very conditions for continued life on Earth is unprecedented; it is the historical novelty that defines the Anthropocene as the age of anthropogenic extinction. As such, it not only forces us to rethink what responsibility calls for. It also calls for a reassessment of the constellation I introduced above, that of the ecologically required and the morally right.

III

In his major work *Respect for Nature*, Paul W. Taylor, one of the pioneers of environmental ethics in the United States, approaches this constellation under the heading "biology and ethics". Taylor points out that there are "two different but equally correct conceptions of human beings: as biological organisms and as moral agents" (2011 [1986]: 47). The question is: what bearing does the one have on the other?

Taylor states that as biologically functioning organisms, our survival requires that we "maintain a certain ecological balance in our relations with other living things"; "we are biologically dependent upon a sound, stable order in the Earth's natural ecosystems" (2011: 48). However, and this is the crux of the matter, "understanding ourselves as biological entities does not provide us with any particular directives as to how we should conduct our lives: our proper role as moral agents is not deducible from the facts about our biological nature" (2011: 49f.).

That our make-up as biological entities *as such* should be able to guide us in matters moral may appear so far-fetched an idea as to merit no serious discussion. But in fact, says Taylor, the claim is frequently made that ecology (defined as a branch of biology) shows us how to live in relation to the natural environment: "Thus it is held that ecological stability and integrity themselves constitute norms for environmental ethics" (2011: 50). Although he doesn't bring the quote, it is clear that the formulation Taylor has in mind is the one for which Aldo Leopold will forever be remembered: "A thing is right when it tends to preserve the integrity, stability, and beauty of the biotic community. It is wrong when it tends otherwise" (1949: 224). Taylor's gloss on this is that "the science of ecology provides us with a model to follow in the domain of environmental ethics" (2011: 51).

For reasons to do with his being a Kantian, albeit one who seeks to make Kant fruitful outside of the anthropocentric framework, Taylor rejects the model where ecology serves as a proto-ethics. His main argument is that the model is not sound "from a logical point of view: it confuses fact and value, 'is' and 'ought'". The ethical question is "How *should* human culture fit into the order of nature?", and this is not a question of "biological fact". Taylor seeks to corroborate his argument by saying that "nothing in ecology can tell us that it is wrong to have a wholly exploitative attitude toward nature" (ibid.).

I think that Taylor's case against the model is much weaker than the case in support of it. This is so because, as explained in Chapter 3, I am a moral realist: I do not accept the claim about a strict separation between fact and value. Quite the contrary, value is part of the furniture of the world, something we encounter *in re*, as objectively there, making demands on us in terms of – precisely – what we ought to do and what we ought to avoid and prevent, such as exploitation of particular manifestations of life. In fact, in the course of his argument Taylor himself (if only unwittingly) subscribes to the realist view, namely when he argues that what is worthy of the respect of all moral agents is the natural entity's having (pursuing) what he rightly calls "a good of its own", a good in the form of flourishing as the peculiar species it (the individual, the specimen) is, a good the realization of which is intrinsic to the entity in question, hence objectively present in it and endowing it with what Taylor refers to as its "inherent worth" (2011: 72). In making the claim that these are the properties found in the entity that merit our respect as moral agents, he is doing precisely what he faults moral realism for: namely pointing to how *the facts about the kind of entity we encounter issue a demand on us* that is normative in the action-guiding sense of morality: a demand on us to act so as to do justice to the entity in question, to support its inherent claim to flourishing in its species-specific fashion. Hence, how we should act, morally speaking, is derived from the peculiar nature (being) of the entity or phenomenon we respond to. Indeed, I would suggest that Kant himself can be said to be a case in point, in that Kant locates value (intrinsic worth, inviolability) *in* the human person, and thereupon derives his ethics' ought from what is deemed the relevant facts. Kant, that is to say, locates what he considers morally speaking of utmost significance, and so what is entitled to respect above everything else, in a specific property (fact) found in the entity, as part of its essential being, in his case the personhood of human individuals, personhood understood as intrinsically linked to capacity for freedom and autonomy, these being the *facts* that exhibit what is *value-able* about such persons, hence decisive for how they ought to be treated.

I am not suggesting that, contrary to his self-understanding, this suffices to make Taylor a moral realist. Nor am I suggesting that the view he dismisses in the version Leopold famously gave it, that ecology serves as a proto-ethics, is sufficient in itself, providing us as moral agents with all the guidance we need for decision-making on the ground. For this position as for its rivals, be it the Kantian one, theory (or model) is one thing, practice another, and so much messier one.

I turn now to the philosopher who in my view offers the most promising approach to the relationship between ecology and ethics, Holmes Rolston. Being a declared moral realist, engaging with Rolston's work will allow me to argue in greater detail why this position is the most promising one within ethical theory given the issues we are looking at. In what follows I will present Rolston's position in his major work *Environmental Ethics* with special regard to the primary topic of this chapter, extinction.

In a passage that may recall the problem of "sacrifice" that Braverman and van Dooren are so preoccupied with, Rolston writes:

> No species can flourish without flourishing individuals who are tokens [examplars, specimen] of the type [species], who exemplify the natural kind. But in the struggle for life, for adaptive fit, many individuals will be losers; their welfare must be sacrificed to predators or competitors. That is not satisfactory to these individuals; their preferences are not satisfied. [...] It may seem unsatisfactory that innocent life has to suffer, and we may first wish for an ethical principle that protects all innocent life. This principle is persuasive in culture, and we do all we can to eliminate human suffering. But ought suffering to continue when humans do or can intervene in nature? That it ought not to continue is a tender sentiment but so remote from the way the world *is* that we must ask whether this is the way the world *ought* to be in a tougher, realistic environmental ethic. A morally satisfactory fit must be a biologically satisfactory

> fit. What *ought to be* is derived from what *is*. In that sense, humans do follow nature axiologically and set our moral standards accordingly. Nature is not a moral agent; we do not imitate nature for interhuman conduct. But nature is a place of satisfactory fitness, and we take that as a criterion for some moral judgments. We endorse a painful good.
> (1988: 58f.)

Given Rolston's reasoning, what does it mean to say – as I have on several occasions – that the capacity for suffering is a sufficient criterion for moral considerability? Let me put it like this. Among the candidates proposed for a property that qualifies for moral protection, the capacity to suffer is better suited than, say, the ability to reason or to speak (to allude to Jeremy Bentham's famous statement). However, we need to situate how that capacity plays out in the environment of which the beings susceptible to suffering are part – suffering as experienced, as lived, as opposed to as theorized within a philosophical ethic. When Taylor states that "all living things have a good of their own, the realization of which is the central goal of their lives (even if that goal is not consciously aimed at)" (2011: 262), I suggest his claim can be reformulated to the effect that to be able to suffer is to be able to *have things matter to oneself*, so that whereas some things promote the realization of the species-specific good, of what it would be like to flourish as the kind of being it is, other things harm it – the point being that the contrast between the two, the positive and the negative, is an objective one, residing in the entity as such, as part and parcel of its very being, as opposed to the contrast being projected onto it, attributed to it by some other agent, say, humans.

Rolston shares Taylor's insistence that value – what demands and deserves protection – is a quality residing in "all living things". In metaethics, this is the standpoint of moral realism, as we saw. More to the point here, it is what the science of ecology informs us: what is value-able is what is constantly brought forth – sustained, regenerated, replenished – in virtue of the interplay between multitudes of living things. In this sense, value is the essence of nature, the modus operandi of the natural world.

Given this perspective, the crucial statement in the above quote from Rolston is this: "A morally satisfactory fit must be a biologically satisfactory fit" (1988: 58). Fit for whom? For which species? From which – or whose – point of view? To connect with the discussion above: that of the individual or that of the species?

Rolston defends a distinct understanding of the relationship between individual and species as applying to, say, non-domestic animals. He approaches this relationship by employing a simple distinction between the domain of culture and the domain of nature. He puts it like this:

> Whatever we say about interhuman duties within culture, humans have no duties…to interrupt the course of wild nature. Nor, when exploiting nature for human interests, have we a strong obligation to reduce

> the suffering below levels found independently of the human presence. Humans are not bound to inflict no innocent suffering. That is contrary to nature. In the clash and interweaving of goods in an ecosystem, pain goes with the defending and capturing of goods that characterize all sentient life. Nothing lives autonomously, not even autotrophs (plants). Everything crowds and competes with neighboring life. One good must spoil another. Heterotrophs (animals) sacrifice other lives for their own. Sentient life both suffer and cause pain. No predator can live without causing pain. [...] The wild animal has no right or welfare claim to have from humans a kinder treatment than in nonhuman nature. What should happen morally in an encounter with nature (distinguished from what happens within culture) is a function of what has happened naturally, Neither in nature nor in culture-nature interactions does need imply duty. So the deer need food, the albatross and turtle need help, but have no claim on us to provide it.
>
> (1988: 59)

To be sure, the clear-cut distinction Rolston makes here, more than 30 years ago, will seem untenable to many today, given that the Anthropocene is the geophysical era where humans are the single most powerful agent on Earth, one whose impact is ubiquitous not only within "culture" but also in the entire natural world. As a consequence, the culture/nature distinction as understood by Rolston becomes problematic: granted that there once was a clear separation/boundary/difference, there is today no demarcation such that "at this point human-made culture comes up against the limits of its operations and their consequences, and henceforth nature takes over, remaining what it is in a state of purity and autonomy vis-a-vis culture". Rather, what we find within what used to be a (more or less) autonomous natural world is the footprint of human activities everywhere.

Even though I accept that the culture-nature distinction is nowadays being questioned in the clear-cut fashion Rolston conceives it, namely as two ontologically distinct and separate domains, I think that the often-made plea for its complete rejection and for doing without it altogether is misguided and risks throwing the baby out with the bathwater. The ontological status of nature in the Anthropocene is a big issue in its own right, one I shall address below.

Recall Rolston's claim that "the albatross and turtle need help, but have no claim on us to provide it" (1988: 59). Does this mean that he would oppose efforts to help the individuals of these endangered species, efforts of the kind described by Michelle Bastian and Thom van Dooren above, with a strong sense of urgency and alarm at that?

The answer is no. In fact, Rolston provides a particularly strong case for fighting extinction of animals like albatross and turtle. That he does so is to do with the role of humans in causing the animals to be at peril: the greater our role in causing them to be at risk, the more their being endangered is a genuine ethical issue in Rolston's sense of moral agency as arising within culture, but not within nature. Rather than Rolston's original distinction between culture and nature being unsuited for the issue, the distinction is *required* to make the case for why extinction is morally wrong.

Rolston proposes the following ethical rule: "Do not cause inordinate suffering, beyond those orders of nature from which the animals were taken. One ought to fit culture into the natural givens, where pain is inseparable from the transfer of values between sentient lives" (1988: 61). So the relevant question is not simply Bentham's "Can they suffer?", but "Is the human-inflicted suffering excessive to natural suffering". On Rolston's view, "ethics does not require us to deny our ecology but to affirm it" (ibid.). What we come to understand, then, is that for Rolston *suffering as such is not an ethical scandal, a wrong to be righted*; likewise, pain as such is not something that ought to be prevented or reduced everywhere encountered and at all costs, as were fighting pain the moral issue and the moral duty par excellence, trumping and overshadowing all other phenomena in nature and being the one thing that moral agents – humans – should always prioritize. Whereas intentional infliction of pain is an evil in culture, the pain that one animal causes another in nature is not an evil, nor is its existence there a moral issue at all. "Pain in nature is situated, instrumental pain", writes Rolston; "it is not pointless in the system, even after it becomes no longer in the interests of the pained individual" (1988: 60). Accordingly, "animals have no right to be removed from their niche; neither have we a duty to remove them" (ibid.), we humans being the only agents who may seek to do so. To hold that suffering as such, pain in itself, is a moral issue in the sense of something that ought *never* to be, that should be prevented or reduced *whenever* possible, would mean – if consistently acted upon – that humans ought to interfere whenever they can when one animal causes pain to another; it would mean to interfere with the interplay between predator and prey that is inseparable from the conditions of life of animals in nature, as opposed to animals in culture, domesticated ones for whose conditions humans are responsible.

The specifically *moral* implication of upholding the distinction between nature and culture as distinct realms, then, is that the pain and suffering occurring between animals is morally neutral, being devoid of blameworthy agents, whereas the pain and suffering inflicted on animals by humans, being moral agents, is in need of justification, on a case to case basis, and hence not *as such* morally prohibited and so always wrong. (This is in keeping with my critique of Gary Steiner's position in Chapter 4.) Having said this, we should be in no doubt about the moral wrongness exhibited in cases where the suffering that animals inflict on one another is a direct consequence of how humans, whether intentionally or not, produce the conditions where that suffering comes to pass, as when, say, pigs attack each other as a consequence of lack of space, neglect, or sheer mistreatment.

Let us return to the key issue raised above: the constellation of the ecologically required and the morally right. As we saw, Rolston's formula is that "a morally satisfactory fit must be a biologically satisfactory fit" (1988: 58). What does it mean to posit biological fit as a framework for moral assessment?

Rolston's proposed formula is a plea for the necessity of proceeding concretely where his philosopher colleagues tend to proceed abstractly – his is not an ethics of the formal and procedural kind championed by Kant and Habermas. For Rolston, the constellation ecology/morality is first and foremost an ontological one, always tied to a particular spatio-temporal location: where the entities belong and play out, the entities whose status in terms of value, rights, and duties philosophers are typically preoccupied within an abstract fashion, focusing on the properties-cum-capacities considered as intrinsic to the individual entity in question, neglecting to specify the extra-individual conditions of its existence, the system (whole) it is part of and cannot exist without.

Importantly, Rolston rejects that ecology is a proto-ethics – i.e., the view we saw Paul Taylor associate with Leopold. Leopold was a trained forester and a philosophical autodidact; his was an ecologically informed, natural sciences based holism, not a well thought-out metaethical position. Rolston is adamant that human conduct is not to be modelled on animal conduct. And we saw that as a rule humans should not interfere with the conduct of animals towards animals. Indeed, humans are very special, even exceptional. Yet in saying so, Rolston means to criticize anthropocentrism, not to commend it; the position he advocates is a holistic biocentrism, informed by the science of ecology but keenly cognizant of the differences between how humans exist in the world and the way animals and other non-human creatures do, these being the very differences that an environmental ethics centred on responsibility needs to take into account.

This may sound like Heidegger, but the point is not to make claims about humans being the only beings who die, who know it and are concerned with it, to recall Derrida's discussion. It is the other way around, meaning: the things that make humans different from all other living creatures are the things that help make us responsible for their plight, for the whole of which they, as well as we, are part, meaning for the future of that multispecies whole – ultimately, for *there being a future for the whole*, to allude to Jonas' formulation. To reject anthropocentrism as we know it and have acted it out in practice in our society, then, is *not* to deny or downplay what is special about humans. Rather, it requires that we recognize the features that make us, in some respects, unique as the very features that bestow on us the moral assignment *par excellence*: responsibility for the biocentric whole.

According to Rolston, humans are in the world ethically, cognitively, and critically like no non-human animal is. In virtue of these features, "the human capacity for a transcending overview of the whole makes us superior and imposes strange duties, those of transcending human interests and linking them up with those of the whole natural Earth" (1988: 72). Moreover, "humans are not important as predators and prey; they play no role in the food chains or in regulating life cycles. They are a late add-on to the system; and their cultural activities (except perhaps for primitive tribes) only degrade the system, if considered biologically and ecologically". Again, our differences from the rest of the living creatures place a demand on us to protect and safeguard them: there is something value-able, value-(re)generating, and value-enriching in their differences from us, a manifestation of value in the world, in the system, that we are obliged to respect irrespective of its instrumental use to us: it is value-able in itself, deserving to be valued because *able to bring forth and sustain value* in the whole natural world. Rolston's way of putting this is to say that "one [specifically, distinctive] human role is to respect the ecosystems they culminate, and not merely to admire and respect themselves, as traditional ethics urges" (ibid.). So there is a sense in which humans are unique and superior, but it is a superiority that is "linked in a feedback loop with the whole" (ibid.).

Humans, then, are *not* unique with regard to being carriers of value, to possessing value. Value is not a human prerogative, is not the creation, projection, or attribution of humans, does not require humans to exist in the world. To say this is of course not to deny that within culture (arts, artefacts, etc.) humans are genuine creators of value, of valuable works and products, including spiritually valuable such. Presently, though, the point is that value is part and parcel of the fabric of the natural world, such that it cannot be adequately described were value absent from the account of its very being. "Some values", writes Rolston, "are already there [in the natural world], discovered, not generated, by the [human] valuer because the first project is really the natural object, nature's project; the principal *projecting* is nature creating formed integrity. Beside this, the human projecting of value is an epiphenomenon" (1988: 117).

Having presented Rolston's understanding of value, we are ready to examine in greater detail his approach to extinction.

Rolston shares the Aristotelian view endorsed by Taylor, that every organism has a good-of-its-kind, a good that it is programmed to defend as a good kind, as worthy of continuation and regeneration. In this sense, everything with a good of its kind is "a good kind" and thereby has value. Hence the living individual, taken as a "point of experience" in the web of interconnected life – and I observed above that the individual is the locus of experience, of having things matter to it, as being good or bad, sustaining existence or threatening it – is per se an intrinsic value, i.e., valuable irrespective of its instrumental value to others (1988: 100).

However, a place has to be made, both biologically and philosophically, for *death* in the system. The truth is: "Without death there can be no life. If nothing much had ever died, nothing much could have ever lived. Even the aging processes that break down life...are goods when incorporated into the system, though they are evils [but not moral evils – AJV] for individual organisms" (1988: 104). Value manifests itself in diverse kinds of experience, a diversity neglected in the hedonist theory of value advocated by Singer and his followers, holding that pain is nature's only disvalue and pleasure its only value, or at least the only value that counts, morally speaking. In the holistic theory Rolston defends, where ethics is "not merely an affair of psychology but of biology", pains and pleasures are part of a larger picture, "derivative from and instrumental to further values at the ecosystemic level, where nature evolves a flourishing community in some indifference to the pains and pleasures of individuals, even though pain and pleasure in the higher forms is a major evolutionary achievement" (1988: 108).

Having argued that the pain and pleasure experienced by individuals are not trump, what is, according to Rolston? Are there duties to species? If there are, what about the moral status of individuals being overridden by the value of a species, be it another species than the one the individuals in question belong to?

These are the questions that came up when we discussed Braverman and van Dooren above, recalling the ethical dilemmas of "the logic of sacrifice".

Rolston acknowledges that there can be duties to species, especially to endangered species; duties that "might supplement or challenge – even override – the duties humans have to persons or to individual animals and plants" (1988: 138). Why can there be such duties? Is it because we are (partly or wholly) responsible for the current situation where so many species are on the brink of extinction? Is it a matter of responsibility for the consequences of a devastation caused by us? If that is Rolston's view, does it really differ from that of an anthropocentric ethics like Kant's or Habermas', or the non-anthropocentric one advocated by Singer? Indeed, what is it that allows us to hold that a species going extinct is *wrong*? Is the answer to be found in our role (responsibility, culpability) as the *agents* causing the harm, or is it found in the *entity* (individuals/species; life form) being harmed? How important is it what precisely helps *cause* its going extinct?

Admittedly, there is something overly textbook-like and simplistic about posing the questions like this. Note in particular the tendency so deeply entrenched in Western ethical theories: to pose the possible answers in terms of either/or. *Contra* this assumption, Rolston sees it as a matter of both-and: yes, the extinction of species is wrong because human activities now help cause such extinction at a pace unprecedented in history. And yet, at the same time *extinction is wrong because it wrongs (violates) the value present in the species going extinct*; consequently, a world where that species is gone would be an impoverished world, one less vibrant and diverse in value. To answer both-and means to highlight the combination of the human-related and the natural world-related aspect of the phenomenon of species extinction. Humans qua agents are the wrongdoers, the species affected in the natural world are the ones being wronged.

IV

I turn now to the issue that came up when we discussed Tom Regan's case for animal rights above: how to assess the relationship between individuals and species, whereby that assessment cannot be purely philosophical (abstract) but needs to engage with the challenges posed by the ongoing extinction of many animal species in today's world.

Rolston notes that a consideration of species offers "a biologically based counterexample to the focus on individuals – typically sentient and usually persons – so characteristic of Western ethics" (1988: 143). Indeed, for the leading ethicists of Western modernity from Hume, Locke and Kant to Rawls and Habermas, the focus on the status of individuals has been taken for granted in the sense that they never seriously considered that the *species*, as distinct from particular individuals, could be endangered. Only with Jonas' book *The Imperative of Responsibility*, first published in Germany in 1979, did the tacit double premise of traditional ethical thought receive questioning: that humanity would persist in the foreseeable future, and – even more tacitly assumed – that the natural world required to support the future existence of humans on Earth would be intact to do so. Small wonder, then, that this tradition is ill-prepared to address a situation where what demands attention is the endangered future of non-human species, possibly undermining that of humanity itself.

The required shift of emphasis from individual to species also reveals the *static* and hence ecologically uninformed way of conceiving the relationship between the two that has prevailed historically. Rolston points out that "in an evolutionary ecosystem, it is not mere individuality that counts; the species is also significant because it is a dynamic life form maintained over time by an informed genetic flow. The individual represents (re-presents) a species in each new generation. It is a token of a type, and the type is more important than the token" (1988: 143).

Ethicists of many stripes will surely seize upon this "more important than": is Rolston advocating that the individual be sacrificed for the sake of the species? But such a reading is not only premature. It misses the point, which is that of dynamism: species are a future-oriented medium of ongoing change (modification, alteration) in a manner the individual (exemplar, specimen) in itself is not and cannot be, existing for a limited time and yet benefiting from the dynamism incorporated in and sustained by the species it represents, a species that for its part depends on the contributions of the individuals that at any specific point in time and space make it up. Rolston stresses the connection between this future-oriented dynamism and the constant (re)generation of value that is the crux of the individual-species nexus: "There is a specific groping for a valued *ought-to-be* beyond what now *is* in any individual. Though species are not moral agents, a biological entity – a kind of value – is here defended. The dignity resides in the dynamic form; the individual inherits this, exemplifies it, and passes it on" (1988: 143).

Rolston's argument for the moral implications of his conception of species is worth quoting in full:

> Because a species lacks moral agency, reflective self-awareness, sentience, or organic individuality, we may be tempted to say that specific-level

> processes cannot count morally. But each ongoing species defends a form of life – on the whole, good things; prolife impulses that have achieved all the planetary richness of life. All ethicists say that in *Homo sapiens* one species has appeared that not only exists but ought to exist. But why say this exclusively of a latecoming, highly developed form? Why not extend this duty more broadly to the other species (though not with equal intensity over them all, in view of varied levels of development)? These kinds too defend their forms of life. We humans are the product of such defenses during long eons past. Only the human species contains moral agents, but perhaps conscience *ought not* be used to exempt every other form of life from consideration, with the resulting paradox that the sole moral species acts only in its collective self-interest toward all the rest.
>
> (1988: 143f.)

Surely the "perhaps" in the last sentence is an understatement. The duty humans, the only moral agents on Earth, have towards other-than-human forms of life and species is not only to do with the prudence of safeguarding the other-than-human conditions of life that the future existence of the human species de facto depends on. It is also a duty that reflects an attitude of gratitude, even awe, if not exactly in Albert Schweitzer's sense.

Though not noted by Rolston, I see a similarity with the mentality of hunters in so-called pre-modern societies committed to an animistic cosmology: the insight that humans are an eminently dependent species, dependent on the thriving of all the "others" that help sustain us into the future, meriting gratitude for the indispensable services they provide to our benefit. *Contra* the hubris that is part and parcel of anthropocentrism as we have practiced it in recent history, and exposing the widespread *denial of the dependence* in question, challenging as this dependence does our culturally entrenched sense of autonomy and entitlement (see Vetlesen 2015, 2021; Orange 2017), Rolston insists that we need to view our situation from a longer-range, objective perspective, one where "systemic nature is valuable intrinsically as a projective system, with humans only one sort of its projects, though perhaps the highest. The system is of value for its capacity to throw forward (pro-ject) all the storied natural history" (1988: 198).

We are rarely encouraged to take this perspective. Doing so may be uncomfortable, but it is first and foremost sobering, and it has the great merit of being true to the biological facts that situate our human existence in time and space. On the longer-range scale Rolston talks about, the fact is that "humans come late and it seems short-sighted and arrogant for such latecomers to say that the system is only of instrumental value for humans, who alone possess intrinsic value, or who 'project' intrinsic value back to nature". Dropping the moderate tone of the "perhaps", Rolston is adamant that "the only fully responsible behavior is to seek an appreciative relationship to the parental environment, which is projecting all this display of value" (1988: 198).

Given this perspective, we are in a position to answer the question "why is extinction wrong" more precisely than I did above.

Rolston's answer combines the biological and the ethical, showing how the former provides the conditions within which the latter must operate. "The main thing wrong", he writes, "is that extinction shuts down the generative processes. The wrong that humans are doing, or allowing to happen through carelessness, is stopping the historical gene flow in which the vitality of life is laid and which…is the same as the flow of natural kinds" (1988: 144). Offering a richer as well as more accurate definition of the phenomenon of extinction than do Braverman and van Dooren, Rolston explains that "every extinction is a kind of *superkilling*: it kills forms (species) beyond individuals; it kills 'essences' beyond 'existences', the 'soul' as well as the 'body'. It kills collectively, not just distributively". This being so, "a duty to a species is more like being responsible to a cause than to a person. It is commitment to an idea (Greek, *idea*, 'form', sometimes a synonym for the Latin *species*). This duty is a categorical imperative to living categories" (1988: 144).

Note that Rolston formulates the duty contained in this imperative without invoking the notion of rights, thus parting company with Singer and Regan. Rolston's position is that we, outside of culture understood as the domain where humans relate to other humans, better dispense with the noun *rights*, since "this concept is not something that attaches to animals in the wild", and that we instead use only "the adjective, *right*, which applies when moral agents encounter nature and find something there judged to be good (appropriate, valuable) before human moral agents appear" (1988: 51). Dispensing with rights, Rolston concentrates on value, value understood as a biologically generated phenomenon, as something factually existing in the world prior to and independent of humans, yet as something that humans, morally speaking, ought to respect *precisely in virtue of its objective quality*. Ethical conduct as incumbent on humans, then, being the only moral agents, is about *the duty to value (respect, protect) the value encountered in the wider world*.

This is my way of putting the synthesis I take Rolston's position to offer between biology on the one hand and ethics on the other, affirming the affinity discussed above between ecology and moral realism, yet without collapsing the one into the other.

To elaborate on the proposed duty, what humans ought to respect is "the living drama, continuing with all its actors" (1988: 145). Amounting to a superkilling, we may say that ethically extinction constitutes a "superwrong". It is the worst that can happen in that it is definitive – irreversible, irreparable – in its effects. Extinction is a matter of "stopping the flow of life"; to kill a species is to "shut down a unique story". Extinction represents an "unnatural end" in that its occurrence is not inevitable; "humans", Rolston observes, "ought not to play the role of murderers" (1988: 145).

That species go extinct is a part of the history of evolution as long back as we are able to trace it (see Barnosky 2009; Kolbert 2014). In speaking of extinction as "superkilling", however, Rolston concentrates on the variant characteristic of the present epoch, the Anthropocene: extinction as human-caused, hence as "unnatural" in Elizabeth Kolbert's sense. Writing a good decade before the term "Anthropocene" was coined, Rolston matter-of-factly remarks that "one form of life has never endangered as many others", adding that "never before has this level of question – superkilling by a superkiller – been deliberately faced" (1988: 145). Precisely because the superkiller species – taking itself to be a superspecies, exceptional in its status and entitlement vis-à-vis all others – is more knowledgeable and have more sophisticated ideas about how present-day actions and events will play out in the form of multiple consequences in the future, that very equal-to-none power comes with a duty that is likewise unique: duties that do not attach simply to individuals or (fellow) persons but are "emerging duties to specific forms of life". If, "in this world of uncertain moral convictions, it makes any sense to claim that one ought not to kill individuals without justification, it makes more sense to claim that one ought not to superkill the species without superjustification" (1988: 146). Surely no such justification exists.

So what follows for assessing the relationship between individuals and species? And given Rolston's emphasis on the importance of species, what happens to individuals?

Clearly, life on Earth cannot exist without species. Nor can it exist without individuals. In reality there are only species as long as there are individual exemplars (specimen); indeed, what happens when a species goes extinct is that its last individuals perish. Even so, there is a difference: a lost individual is always reproducible; a lost species is never reproducible. What follows from this difference?

As we have seen on several occasions, the dominant ethical theories are premised on individualism – ontologically, methodologically, and normatively – and therefore only allow duty to individuals, not to a collection (collective entity) such as species. Suffice it to mention here Peter Singer's position as a case in point, cited by Rolston: "Species as such are not conscious entities and so do not have interests above and beyond the interests of the individual animals that are members of the species" (Singer 1979: 23). However, even if we grant that such things as beliefs, desires, conscious awareness, preferences, interests, and (even) rights pertain to *individuals*, not to species, this fact does not mean that only individuals, and not species, are suited as objects of our duties. Why?

It is because what matters ethically (normatively) cannot be separated from what matters ecologically and biologically (factually). What we value, what merits that we value it, in the form of observing duties towards it, is value, what is value-able; that is to say, the forms in and through which value exists in the world, and continues to do so into the future. Absent species, absent the existence of value in the world, comprised, sustained, and regenerated as it is by way of species. As for individuals, their belonging to a species, instantiating it, exhibiting it, helping reproduce it, is what facilitates – alongside a multitude of *other* species, other life forms – their having value, their being carriers of a value that *transcends* each one of them qua individuals, stretches, and lasts beyond their limited life cycle, the particularities of their concrete spatio-temporal trajectory. Individuals partake of an ongoing story of value, a never-pausing enactment of value in the world, that is bigger than each individual in itself.

Unexpectedly, Rolston's unconventional view of the relationship between individuals and species has met with criticism. In *The Ethics of Species*, Ronald Sandler writes that "the key mistake in Rolston's argument for the inherent worth of species is the inversion of which entity – the species or the individual - is primary in survival and selection processes" (2012: 38). In particular, Sandler takes issue with Rolston's statement that "the species too has its integrity, its individuality; and it is more important to protect this than to protect individual integrity. The appropriate survival unit is the appropriate level of moral concern" (Rolston quoted in Sandler 2012: 38). Sandler seeks to counter Rolston's view by pointing out (as does Tom Regan) that "the 'survival unit' is not species, but individual organisms. Individual organisms are what live, die, and reproduce (or fail to do so)" (ibid.). This, we should observe, is in keeping with Singer's assertion that "all our duties must be to sentient beings", taken as so many individuals (see Rolston 1988: 146).

Sandler's criticism does not stand up to closer scrutiny. To begin with, the "inversion" for which he faults Rolston seems to be based on the assumption so typical in dominant ethical theories: that the possession of inherent worth is a matter of either/or: either the species or the individual. Second, there is something mechanic, not organic, about the way Sandler's makes the distinction between species and individuals. In setting them apart as separate entities with separate properties (i.e., only individuals have interests and integrity), Sandler fails to do justice to their being mutually supportive and complementary in their actual existence. In this, Sandler conspicuously comes across as the analytic philosopher of the two, whereas Rolston's approach benefits from his background in physics and the natural sciences.

In particular, Sandler's criticism is weakened by neglecting a crucial element of Rolston's view, namely: "The locus of intrinsic value – the value that is really defended over generations – seems as much in the form of life, the species, as in the individuals, since the individuals are genetically impelled to sacrifice themselves in the interests of reproducing their kind" (1988: 148). Several points are important here. First, in stating that the locus of intrinsic value "seems as much in the form of life, the species, as in the individuals", Rolston's position is conceived in terms of both-and, not either/or. He is not denying intrinsic value in individuals; he is arguing for it in species *as well*. And the value upheld in species, thanks to everything and everyone that contributes to the transfer of that value through time and space, is not "in addition" to the value quasi-separately found in individuals; rather it is *co-constitutive* of it, helping facilitate and sustain it through time and space.

Second, the issue of status in terms of value that Sandler is concerned with is not a conceptual one than can be settled without taking into account the actual life conditions of the entities in question. It is fine to hold, as does Sandler, that every individual possesses inherent worth. But the question is: what is required for a particular living individual (say, a lion) to realize and *maintain* that worth through its life cycle? The worth that is "in" the individual lion cannot be realized *by* that individual as such, apart from other lions, i.e., other individuals of the same species, nor without its inter-connections and inter-dependencies with other species, maintained within a shared habitat and a wider ecological network. To save the lion, you need to save the jungle: the relevant unit, morally as well as ecologically, is the lion-in-the-jungle. Individualism of the sort Sandler is committed to simply doesn't work in the setting of the natural world where the individual lions live their lives and manifest their value, and where the contributions of its many others are indispensable for the very constitution, as well as sustenance, of that value.

Finally, there is Sandler's individualism in the sense that "an entity's moral standing, when it comes to moral worth, is determined by the properties or capacities of the entity itself", qua individual at that (2012: 41). Well, *it is not* – not factually, as explained above. So, when Sandler makes his plea for "a capacities-oriented approach", where capacities are properties of individuals only, "rather than a species membership-based approach" (ibid.), he is again committing the error of either/or thinking. The deeper problem, I suspect, is that Sandler's reasoning is based on human-to-human relationships as the proto-model for ethics in general and the issue of value in particular, even though he wishes to go beyond anthropocentrism á la Kant in recognizing what Kant blatantly denies: inherent worth in animals.

Let me put what is at stake here like this. To hold that the worth of a *human* individual derives from and is dependent on his or her belonging to a specific group (collective) of humans, be it defined in terms of sex, ethnicity, religion, or class (not to mention race), would be ethically fatal, perhaps also literally fatal for the individual in question, depending on the circumstances. Thus the sanctity of human persons, their inviolability (Kant), has to be *individual*-based not collective-membership-based. Historical cases of genocide demonstrate the danger involved once the worth of humans is made a function of, contingent upon, *which group they belong to*, where membership to one such (say, Aryans, or whites) is tantamount to being "worthy", and membership to others (say, Jews, or coloured) is tantamount to being "unworthy" (see Vetlesen 2005).

In saying this, I do not deny the deep existential significance that belonging to a specific collective (group) of humans has for us humans. I am only saying that, for all its importance, such belongingness to a whole of which I am a part (member) must have an element of freedom, of choice, namely as to *which collective* I wish to belong to and to align myself with it terms of my values and projects in life. This must be an optional matter – it is indeed valuable in itself that it is so – and cannot be a condition of our having moral standing as such. There is, then, a sense in which Kant got it right when stressing the strictly individual nature of the inviolability of human persons – this inviolability is found, and has be respected, in humans in a stricter sense than, say, in individual animals, since their belongingness to and alliances with a collective is never optional, a matter of choice and freedom, the way it is for human individuals: we are free to negotiate our relationships with our group-belongingness in ways not available to, or necessary, for animals. Let me reiterate here what I said above about what is special about us as moral agents – and we are the only such – namely that we have a responsibility for cultivating our capacities for responsibility. The gift of responsibility comes with the task of safeguarding it, a specifically human task shared in practice by each individual and his or her institutions, to put it in the spirit of Hegel's *Philosophy of Right*.

This may appear to be a distraction from the topic under discussion, but it is not. One of the primary tasks for a non-anthropocentric ethics is to acknowledge that there are morally important differences between humans and non-humans. I have just paid attention to one such, of paramount importance. While Rolston acknowledges that there are such differences, he should have engaged more deeply with their implications. We certainly need to have their relevance in mind when he states, as we saw above, that "the individuals are genetically impelled to sacrifice themselves in the interests of reproducing their kind", adding that "the life that the individual has is something passing through the individual as much as something it intrinsically possesses. The individual is subordinate to the species, not the other way around" (1988: 149).

Would we accept this statement if it were meant to apply to humans? I for one would not. As individuals we are not ready to "sacrifice" ourselves in the interests of reproducing our kind, at least not if that sacrifice entails that we willingly give up our lives – prematurely – for the sake of our species. Often the rhetoric of sacrificing some for the allegedly higher worth and bigger cause of a collective of some sort has prepared for and helped legitimize mass killing. This being so, we do not see ourselves as "subordinate to the species", nor should we.

However, the context in which Rolston makes these statements is his chapter on endangered species, where his focus is throughout on non-human such. The sacrifice and subordination referred to is descriptive, articulating the modus operandi of non-human species and life forms in the natural world, which – for good or bad – Rolston as we recall distinguishes from the human-made world of culture. His focus on non-human life is a given premise in his discussion, as is the tacit assumption that in talking about endangered species, *Homo sapiens* is not one of them. Rolston clarifies the distinction between an individualistic framework and what he advocates as a more "comprehensive" one, truer to the facts as he understands them and in keeping with his formula that "a morally satisfactory fit must be a biologically satisfactory fit" (1988: 58). If an individualist will hold that species-level phenomena such as vitality in population and reproduction of a life form are only epiphenomena, by-products of aggregated individuals in their interrelationships, Rolston's comprehensive – holistic – approach depicts "the species itself as a kind of individual, historic lineage over time"; to be sure, "we want individuality, too, but also at the species level; individuality within community, the one as real as the other" (1988: 149). Rolston is perfectly aware that he will meet the objection (recall Sandler) that a species does not act on needs or with interests because there is no "it" to act, that is, only individuals can do so. But Rolston is adamant that "the specific type, no less than individual tokens, is an 'it'. A specific form of life urges survival of 'its' kind, defends 'its' life form. The 'it' is a historic process with vital individuality, though it is not a single organism. This helps us to see that the telos of a species is not a fixed end but rather one evolving over the propagation of the gene linkage, an endless end" (1988: 149f.).

I conclude that an approach that recognizes value as intrinsic to species *as well* as to distinct individuals is more suited to address extinction than one that recognizes value only in individuals. Even if the death of the last individual exemplar is the local event that means that the species goes extinct, the *kind of loss* that extinction represents is best conceived on the level of species: that the species is forever lost from planet Earth is the deeper and enduring *meaning* of the death of the last rhino, capturing the magnitude of the loss, the fact that its importance transcends whatever meaning it may hold for the single individual in question. That this is so follows from Rolston's claim that value is something objective, not merely subjective. The loss implied in extinction does not hinge on the way in which the individual animal relates to it owing to its capacities for experience, for pain and pleasure; it is neither derivative from nor reducible to such subjective experience.

Therefore, to evaluate the entire biological world on the basis of sentience – as does Singer, probably the most influential individualist and subjectivist within environmental ethics – is a category mistake as far as Rolston is concerned, since from a biologically attuned point of view this position mistakes a part for the whole; and conceptually it can only get at the whole by aggregating the individual-based input, missing out on its qualitative distinctness and thus on what is at stake in the event of extinction (a phenomenon not part of Singer's original agenda, it must be said). Rolston readily accepts that "the highest value attained in the system is lofty individuality with its subjectivity, present in vertebrates, mammals, primates, and preeminently in [human] persons" (1988: 191). But again, such products are not the sole loci of value, "concentrate value though they do". Rolston's synthesis of value objectivism (realism) and holism boils down to this: "Even the most valuable of the parts is of less value than the whole. The objective, systemic process is an overriding value, not because it is indifferent to individuals but because the process is both prior to and productive of individuality" (ibid.).

V

In their introduction to the anthology *Extinction Studies*, Deborah Bird Rose, Thom van Dooren, and Mathew Chrulew make a comment on Elizabeth Kolbert's critically acclaimed *The Sixth Extinction* that I find relevant in light of the above discussion of Rolston. "Humans", they write, "really appear in Kolbert's pages in only two forms: that of the (often heroic) conservationist struggling to hold onto disappearing species, and that of an amorphous and monotonal 'threat'" (2017: 6). What is missing is "a detailed discussion of the specific political, economic, and cultural forms of human organization most responsible for any given extinction", a discussion that would need to recognize "radical inequity and highly differentiated positioning" as the name of the game (ibid.).

This criticism applies to Rolston no less than to Kolbert. Not only does Rolston fail to provide an account of the societal forces that precipitate the degradation of the natural world and the urgent spectacle of extinction. He also has very little to say about the sorts of conflicts that are bound to be generated given his distinction between entities that exhibit comparatively "higher" and "lower" value, with human persons placed at the top.

To be sure, there are plenty of books addressing the various societal agencies of destruction that Rose et al. are pointing to, Naomi Klein's *This Changes Everything: Capitalism vs. the Climate* being perhaps the most influential one. Despite his sparse attention to the political agents and economic structures that Klein rightly concentrates on, two aspects of Rolston's approach are relevant here.

The first is to do with what Rolston says about humans. As indicated, I find it a strength that in building his case for a non-anthropocentric ethics, he emphasizes the ways in which humans are truly exceptional. In fact, and despite my observation about the lack of a more comprehensive account, I think that the current human-caused extinction of so many species is especially well suited to Rolston's approach.

Rolston's central claim is that the superiority of humans as compared with all other beings is what grounds our *responsibility* to all those "others", generating our duties to the natural world, geared to respecting the value found there, constantly realized there as our never-suspended, never-ending *conditio sine qua non*. Importantly, *this value is not of our making*; it precedes our existence on Earth, late add-on that we are, even though we may rightly be held to represent a peak value in the history of evolution in terms of subjectivity and individuality. This natural prehistory of ours obliges us to be humble and thankful to the life forms indispensable for our emergence as a species in the past and for our persistence into the future, humanity's dependence on other life forms being a given of our existence that will never cease or be overcome. Owing to our advanced capacities for understanding and imagination, capacities that allow us to abstract from our given situation and surroundings in ways not matched by any other species that we know of, we have an obligation to make good on those exceptional capacities: to exercise them with regard to our many "others" and not only with regard to ourselves. Notice here how sharply Rolston's portrait of human exceptionalism and its implications differs from the one championed from Plato over Descartes and Kant to Habermas and Rawls, all of whom highlight human rationality and, as part of that, the mind over the body, whereby the mind is taken to define the human, and the body to define non-human animals.

If the capacities Rolston foregrounds appear overly intellectual too, where he differs from the canon is in holding that the point of our possessing them like no others do is to *direct them at others*, at entities that (partly or completely) are deficient in those capacities but to be respected as value-able nonetheless. This is in keeping with my proposed formula that *not all moral addressees are moral agents*, nor need they be. The crux of responsibility as only humans can exercise it is that in doing so we do not restrict it to humans.

It might be tempting to say that this is the cost with which these capacities are given to us. But "cost" is the wrong word, born of a wrongheaded approach. In truth it is an enrichment, giving the capacities an ethical task that takes our commitments beyond ourselves, directing them to the biocentric whole of which we are part. Humans, writes Rolston, "can be 'let in on' more value than any other kind of life"; "humans are of capstone value because they are capstone evaluators", borne out in "our capacity to *see others*, to oversee a world". Rolston's ethics, then, calls for "seeing nonhumans, for seeing the biosphere, the Earth, ecosystem communities, fauna, flora, natural kinds that cannot say 'I' but in which there is formed integrity, objective value independent of subjective value", to see "*without* as well as *within*" (1988: 339f.). Thus we can achieve a comprehensive kind of altruism, recognizing the claims of fauna, flora, ecosystems, and landscapes; this ultimate altruism that puts a premium on transcending us, our preoccupation with our own species, is the true human genius. This *subjective* capacity, this capacity of the kind of subjects that we are, to be *objective* in appreciating world objects is a "superiority worthy of extra valuing" (1988: 341). I take it that it is so for the reason we saw Hans Jonas advocate above: if there is such a thing as a responsibility among us humans to sustain our existence qua species, it is to do with our role in making sure that there will continue to be a world where the myriad non-human addressees of that responsibility persist and thrive. Our specialness is not for our sake but for others.

VI

In her important book *What is Nature?* the British philosopher Kate Soper contends that "since any eco-politics, however dismissive of the superiority of *homo sapiens* over other species, accords humanity responsibilities for nature, it presumes the possession by human beings of attributes that set them apart from all other forms of life". All ecological injunctions, she continues, are "clearly rooted in the idea of human distinctness. For insofar as the appeal is to humanity to alter its ways, it presupposes our possession of capacities which are singled out from other living creatures and inorganic matter" (1995: 160, 40).

Ever since environmental ethics emerged as a distinctive field in the 1970s, the notion of human distinctiveness has tended to be looked upon with suspicion – understandably so, in light of its role in justifying the exploitation of everything non-human for human purposes (Crist 2019). Again, Rolston's twist on this notion, turning the responsibility it entails outward instead of inward, is a step forward, showing that a non-anthropocentric ethics needs to hold on to what makes humans special, not to downplay or deny it – nor, for that matter, to conceive human impact on nature as always, per definition, negative and harmful.

There is a catch, however. It has to do with the way in which humans are special in that we have one foot in each of two distinct domains, as it were: culture and nature in Rolston's account. In light of this distinction, our being moral agents endowed with responsibility for all sorts of "others" is an aspect of our being *cultural* beings, not only natural ones.

This may sound like abstract metaphysics, and so beside the point when what we want to illuminate is human-caused extinction. But in fact the distinction is crucial to recognizing the distinctness of extinction qua anthropogenic that we are after. In what follows, I shall leave behind Rolston's version of a distinction between culture and nature and concentrate on more recent proposals, more suited to the present urgency of the topic of extinction.

Soper observes that "green" thinkers have tended to reject dualism as inherently eco-unfriendly. Their reasons for doing so with regard to the Cartesian legacy (humans vs. animals, mind vs. body, mind vs. matter) are well known and shall not be restated here. For Soper, though, a certain kind of dualism provides the *foundation* for an environmental ethics. Why? Because "there can be no ecological prescription that does not presuppose a demarcation between humanity and nature" (1995: 160). There are two aspects to this. The first is the one just discussed, concerning how humans are special, absent which "it would make no more sense to call upon humans to desist from 'destroying' nature than to call upon cats to stop killing birds" (ibid.). Consider this the necessity of differentiating humans from other organic and inorganic forms of being in order to make sense of human agency in general and responsibility in particular.

The second aspect is the one we now need to concentrate on. Having established what humans are, what is nature? Soper defends what she calls a realist concept, equivalent I suggest to the one promoted by Rolston: nature, she explains, refers to "the structures, processes and causal powers that are constantly operative within the physical world, that provide the objects of the study of the natural sciences, and condition the forms of human intervention in biology or interaction with the environment. It is the nature to whose laws we are always subject, even as we harness them to human purposes, and whose processes we can neither escape nor destroy" (1995: 155). Soper sees this realist notion of nature as essential to the coherence of ecological politics. But she hastens to add that accepting it is not in itself to take up a normative position towards nature, specifying what attitudes we should adopt towards it. Even so, nature in the realist sense "sets certain limits on what we can do, or even try to do, and we must observe these on pain either of looking very foolish or else perishing in the effort to transcend them" (1995: 159). Though real and to be ignored only at our peril, the exact nature of the limits set by nature is a complex topic that Earth scientists now struggle to fully understand, conceived as so many "tipping points". But the bottom line is that there are some "conditions in nature" that must be observed if we are to avoid calamities and to be able to preserve nature as a set of resources, or better – with Rolston – the natural world understood as the site of ongoing bringing forth of value. However, Soper submits, "we shall not 'destroy' nature at this level even if we fail to observe the limits in question". Why is that? Nature at this level, says Soper, is "indifferent to our choices, will persist in the midst of environmental destruction, and will outlast the death of all planetary life" (ibid.).

In my view, the realist notion of nature Soper endorses, like Rolston before her, is far better suited to address what happens with nature in the Anthropocene than the influential constructivism championed in posthumanism á la Bruno Latour, subject to extensive critique in my *Cosmologies of the Anthropocene*. For the sake of the important contrast it makes with the realist ontology I defend, suffice it to say here that Latour's view is that "to believe that the distinction between humans and nonhumans and the difference between culture and nature...describe anything at all about the real world amounts to taking an abstraction for a description" (2017: 58). What Latour professes to aim at is nothing short of "blurring the distinction between nature and society *durably*, so that we shall never have to go back to two distinct sets" (2004: 36). This being so, Latour wants to go beyond the no longer subversive-sounding assertion that nature is merely a cultural construct or that wilderness is nowhere to be found in today's world, both firmly installed as the intellectual, and to a growing extent also political, *Zeitgeist*. For him and his numerous followers, to hold that "everything is connected" is to unreservedly, and without regret, embrace the idea that *as a matter of fact the entirety of nature has been anthropized*. To protest this state of affairs, be it philosophically or politically, is pointless, making a fool of oneself. Instead of leaving alone a so-called nature that doesn't exist, we should learn to "intervene even more", since "the environment is exactly what should be even more managed, taken up, cared for, stewarded, in brief, integrated and internalized in the very fabric of the polity" (Latour cited in Neyrat 2019: 92). So it is not only that Latour wants to throw out, dustbin-of-history like, the realist notion that what is wild exists as an expression of autonomy of that which is more-than-human, beyond anthropogenic mastery and control (as argued by Rolston and Soper). It is also that he sees no reason whatsoever to lament the collapse of nature into society, the other-than-human into the human, egging Prometheanism and its current entourage of techno-optimists and techno-capitalists on instead of fighting it and bringing it to a halt. Notwithstanding the avowed radicalness and the air of subversion of this rhetoric, it is, in fact, as Marx-oriented critics such as Alf Hornborg (2019) and Andreas Malm (2018) point out, an exercise in critique dismantling and disarming itself completely, and not merely running out of steam.

Returning to Soper's perspective, the "elasticity" she recognizes in nature doubtless throws up questions that are truly challenging, especially so for a realist conception.

What, more exactly, does the elasticity consist in? Soper, employing the term almost 30 years ago, does not elaborate. Today, following the emergence of the interdisciplinary field of Earth Science, the term draws attention to phenomena like disruption, resilience, and – last but not least – tipping points such as defy conventional linear causality (see Hamilton 2017). Historically novel, these phenomena attest to the current impact of humanity on planet Earth, impact of such a magnitude as to raise questions about the extent to which that impact *alters* life on Earth, so that nature is no longer what it used to be.

Today one can hardly overestimate the kind of change *in* nature brought about by human activities worldwide. Recall how strongly Bill McKibben put his thesis more than 30 years ago in *The End of Nature*: "By changing the weather, we make every spot on earth man-made and artificial. We have deprived nature of its independence, and that is fatal to its meaning. Nature's independence *is* its meaning; without it there is nothing but us" (1989: 60f.). Though suggestive, defining what nature is in terms of its complete independence from humans, undisturbed in its original purity and autonomy, comes with a price: it is as if humans have no appropriate or "natural" place in nature, as if nature, perfect and sufficient unto itself, can only be harmed and eventually destroyed *qua* nature the moment humankind started to make its mark within it – an impression reinforced when McKibben describes encountering other human persons while "in" nature as unavoidably spoiling the experience he is seeking. This becomes especially problematic in that McKibben explicitly laments our present-day *separation* – distance, alienation – from nature. He thereby distorts our belongingness to it and our sense that, at a profound level that helps define who *we* are, we are part of nature, not apart from it. As Emma Marris points out, we must avoid conceiving humans as somehow "unnatural" and by definition "destroyers of nature" (2021: 254).

If the era of the Anthropocene is indeed the era of extinction, it is not because nature has been deprived of the independence McKibben ascribes to it. There is more at stake than the loss of such purity, if it has ever existed. To address extinction in its current form is to confront the fact that native ecosystems no longer exist, given that today there is no spot on Earth that has not been influenced by human activities, indirectly and at the receiving end if not directly, as in the case of animals who have never even been in direct contact with humans, such as polar bears containing PCB and other pesticides and pollutants emitted far away, or albatrosses whose stomachs are filled with plastic.

What to do, then? Should we abandon the idea that there is such a thing as "wilderness", considering it as – precisely – an idea (hence the scare quotes), as an increasing number of scholars now argue? Or should we take the more complex view that, aside from being ideologically charged qua cultural construct, buying into nostalgia for a lost Eden, the strongest case for saying goodbye to "wilderness" is the empirical one verified by the hard sciences: namely that it is gone in the natural world?

Notwithstanding the plausibility of the case thus made, in *Abundant Earth* Eileen Crist urges that we stick to the concept, arguing that "dispensing with wilderness [on the grounds cited] is like throwing the baby out with the bathwater" (2019: 5). Why? Because "to discard the conceptual vessel of wilderness – as referent to the natural world and as an idea in the human imagination – is to forsake nature's original blueprint, which lucidly reflects back to the impoverishment that limitless human expansionism has produced. To forsake nature's original blueprint of prodigious creativity and abundance supports the allowance of its banishment in both reality and memory" (ibid.).

Crist's argument is in full agreement with my own position as well as Rolston's realism and his distinction between culture and nature, implying as we saw that humans belong within *both* domains in a way non-human animals do not, the "cultural" one being what grounds moral agency and responsibility as uniquely *human* features. That is the philosophical and conceptual aspect of the agreement with Crist's view. But the historical-ecological aspect is equally important. Crist is not in any way contesting or downplaying the reality of ongoing degradation, decimation, and destruction of nature. It is precisely in order to assess the causes as well as the consequences of that spoiling of nature, *in* and *for* the nature so disastrously affected, and to do so in a genuinely *critical* manner with regard to the societal forces driving it, that she insists we retain the concept of wilderness, without scare quotes at that.

There are three aspects to Crist's argument. The first two can be separated analytically, like I did, but the crucial insight concerns their being two sides of the same coin. Crist explains:

> As the primal manifestation of nature, wilderness contains its most biodiverse, fecund, creative, and dynamic expressions. To dispense with the referential vehicle for this manifestation of the natural world – as

> a concept both for the real world and the human imagination of that world – is equivalent to jettisoning the original blueprint of nature. This jettisoning, in turn, amounts to doing away with the conceptual vehicle that lucidly reflects the diminishment and destruction effected by limitless human expansionism. To discard the concept of wilderness, as certain strands of environmental thought propose, is to forgo a contrasting baseline reality that reveals the full picture of civilization's onslaught.
> (2019: 114)

The third aspect of Crist's view follows from seeing the conceptual and the empirical as interconnected: to dismiss wilderness as but a modern (Western) cultural construct and to deny the primordial reality of wilderness paves the way for *naturalizing* the human impact and so *normalizing* nature's domination by humans, as were there two eternal truths to be learned here: that humans have *always* dominated nature with whatever means available and that nature has *always* been subject to this domination.

As opposed to this historically uninformed critique, the position with critical bite is to defend wilderness the way Crist proposes: namely understood as "nature's self-creation free from constant revamping, using, and exploitation", is the "core node of resistance against the natural world's domination". Only by holding on to the original ontology of the biosphere as "a vast biodiverse wilderness" will we be able to reveal – through contrast – "how human domination obliterates life's richness" (2019: 115f.). The task of capturing the comprehensiveness of the current onslaught on nature cannot do without *a comparison between what nature was like before that onslaught and what it is changing into in our present era.* Decimation to the point of extinction has become a "normal" occurrence only in the last part of the twentieth century, termed the "Great Acceleration", becoming global in impact and speeding up in particular since the 1970s (see Angus 2016).

The staggering pace and worldwide reach of destruction is doubtless a reason why so many scholars, especially in the humanities and social sciences, now subscribe to the notion that "native ecosystems no longer exist" (Merchant 2016: 154) and that wilderness has become an idea out of touch with reality. Crist warns against being so carried away by ongoing degradation as to conclude that there is nothing – literally, no place on Earth – untouched by human impact; that the entire natural world is being wrecked, so that nothing original or pristine is left anywhere. This is the reasoning that throws the baby out with the bathwater.

For a start, we should not be committed to defining wilderness as 100-percent pristine. Such a definition is spurious, a scientific non-starter, suitable only to ideological distortion. In our time, writes Crist, wilderness has a more robust and pragmatic meaning: "it consists in extensive natural areas that are mostly undisturbed by civilized human activities; many such regions remain in the world, and more can be restored" (2019: 116). Hence the emergence of projects of "rewilding" so as to facilitate the reintroduction of particular species of animals to places where they used to be prolific, such as wolves in Italy and Spain and sea eagles along the coast of Scotland (see Monbiot 2013). To retain the notion of wild nature conceptually so as to help intensify efforts to preserve what remains of it "out there" while there is still time, is not to be committed to a fiction about some "original" and ahistorical state of nature and the species found, as if everything to do with change and dynamism were a product of human "interference" instead of properties intrinsic to nature and what Crist speaks of as "the original ontology of the biosphere". She quotes Rolston's statement that a truly "sophisticated concept of wilderness preservation aims to perpetuate the integrity of evolutionary and ecological processes" (2019: 117). This is an important point, showing that if there is such a thing as a static and ecologically uninformed position, it is that of the detractors of the wilderness view, not its defenders: what is being disturbed, spoiled, undermined, damaged, disrupted – so many words for harmful anthropogenic impact – is *precisely* nature's intrinsic dynamism and creativeness, its inbuilt capacities for facilitating and sustaining flourishing in myriad life forms and species. These are the qualities Rolston has in mind when he says that humans confront a "projective nature", one "restlessly full of projects – stars, comets, planets, moons, and also rocks, crystals, rivers, canyons, seas" (1988: 197).

Shifting the perspective from the ecological to the cultural and historical, Emma Marris is right to point out that "wilderness" is sometimes used in a way that "perpetuates the colonial myth that Indigenous people had no agency and could not modify, manage, or influence the landscape around them"; thus we should beware of, say, measuring the "ecological integrity" of North American landscapes by comparing their current state to the "natural range of variability", typically defined as "whatever it looked like in the three or four hundred years just before Europeans showed up". Indeed, Marris goes on to observe, "Indigenous land management was in many places creating those 'natural' states"; and in the cases "where colonizers saw this management in action, they often failed to recognize it, because of its different forms and because of their preconceived notions about the Native people" (2021: 62f.).

Marris' warning against mistaking "wilderness" for "the places where colonialism left the threes standing" (2021: 160) is well taken, as are her observations just quoted. Yet in my view they do not weaken Crist's case for holding on to wilderness in her well-considered sense. Crist is well aware that there was never a situation where existing groups of humans did not *somehow* impact on nature. The important point to make, though, is to do with the ways in which that impact is nowadays dramatically reinforced, so as to alter nature to a degree that is truly unprecedented in the history of humans on Earth.

But, it will be objected, doesn't the sheer comprehensiveness of the present onslaught against nature force us to give up the ontological division between culture and nature that we saw Rolston endorse? Isn't holding on to that division at odds with the pervasiveness with which culture (i.e., human-initiated activities and their products) now interferes with nature, with what it used to be "in itself"?

It is indeed a well-established fact that the human and the non-human are more intertwined than in any previous epoch, that the two blend, interact, and interconnect to an extent never seen before – a development whose implications was the topic of sociologist Ulrich Beck's path-breaking *Risk Society*, aptly published shortly after the accident at the Chernobyl nuclear plant in April 1986. This may tempt us to treat the distinction not as ontological in some presumably timeless, inalterable sense but instead as eminently concrete, illustrated, for example, by how various animals that until recently lived apart from humans no longer do so: foxes, polar bears, and seagulls become "urbanized", forcibly so since the forest is gone, the sea ice is melting, and the fjord devoid of marine life nutrition – prompting the seagulls to come to my garden, looking for something to eat. This being the sort of change imposed on an increasing number of animals around the world, it is more apt to talk about the animals per force moving place from their old natural habitat to their new urban one than about their being affected by how nature and culture blend and become blurred (i.e., the "hybridity" that so fascinates posthumanist theorists) where they were once ontologically distinct. There is something odd about these animals entering urban areas, relocating so as to stay there (here) for the foreseeable future, something out of joint that brings to mind Freud's notion of the uncanny ("das Unheimliche"), a sense that "there is something where there should be nothing, and nothing where there should be something"; conspicuous presences in combination with conspicuous absences. Whereas for us it's the animals that change their locations in what we experience as uncanny and erratic ways, *for them* it's we humans who do so – it being our expansion that enforces their relocation.

Take the case of tigers. As Crist notes, in 1900 there were an estimated 100,000 tigers in the wild, worldwide, consisting in seven subspecies. Today there are approximately 4,000 individuals left in the wild, and four of the seven subspecies have gone extinct. In other words, 96 percent of tigers have been lost in 120 years. Such a substantial quantitative change in the history of tigers brings with it a change in the way children perceive the qualities peculiar to such an iconic animal. When I grew up, pictures of tigers would elicit a sense of awe and admiration for their superiority, being majestic animals an encounter with which a human person would stand little chance of surviving, and so recalling Plumwood's experience with the crocodile. Nowadays pictures of iconic animals like the tiger and the polar bear in fact elicit a markedly different reaction among children in particular as compared with a few decades ago. Why? It's as if these animals' image, based on their fabulous physical power and agility, their being fearful and admirable alike because so majestic, so sovereign in their independence vis-à-vis us humans, now gives way to an altogether different set of associations, so that they appear *weak rather than strong*, frail rather than sovereign. For most children in Western countries today, the polar bear, say, is the animal caught in the famous photo by Arne Nævra at Svalbard, going viral 20 years ago: a big bear alone on a small ice sheet, struggling to hold onto it and surrounded by open sea, suggesting that this is an animal struggling to find food and to take care of itself and its cubs, facing an increasingly precarious existence where it is forced to give up hunting seals, forced to go on land to search for birds' eggs instead, and if that fails, forced to roam the closest human settlement, driven, for the sake of survival, ever closer to the one species behind the destruction of its habitat. The same process involving ever-shrinking habitats forces the last remaining tigers in India to enter villages where they risk being killed for coming too close to people fearful of their presence, not to mention the risks involved for animals and humans alike when they come into such proximity in wet markets, enabling the transfer of deadly virus from animals to humans, exemplified by the COVID-19 pandemic that originated in Wuhan, China, in December 2019 (see Shah 2017: 2020).

The complaint is often made that in giving pride of place to animals like the tiger and the polar bear, their long-standing status as iconic wild animals in the cultural imaginary of Western modernity means that the large majority of the animals now threatened with extinction are left in the dark, attracting no media attraction, no celebrity status like that of the lion Cecil, world famous for being shot by a rich dentist, duly condemned for his heinous crime by global consensus. It is tempting to say that what we see here is nothing new under the sun, nothing that we haven't seen many times before, and involving human persons at that, not individual animals. Routinely, as a matter of fact, on average 10,000 children die from "easily preventable causes" every day; but as long as they have no name, no face, they fail to attract the attention their plight merits, morally speaking; they fail, that is, to elicit the Levinasian appeal "thou shalt not kill!" (see Boltanski 1999). For every article about the situation of, say, vultures in India, there will be hundreds about that of the Indian tiger or the Sumatran rhino, of which 100 remain in the wild (Braverman 2015: 55). Yet the plight of the vultures is no less critical. As van Dooren reports, today approximately 97 percent of the three main species of vulture in India are gone, in recent years largely as a result of the birds being unintentionally poisoned by a drug called diclofenac, given to cattle whose carcasses vultures eat. In vulture bodies, van Dooren explains, diclofenac causes "painful swelling, inflammation, and eventually kidney failure and death" (2014: 47). Despite tireless conservation efforts, whether there will ever again be a sustainable population of these predator birds in the wild is an open question. To be sure, vultures are not an iconic species like tigers and lions, though it must be said that being iconic is often a mixed blessing – think of what motivated the dentist to seek out and kill Cecil. Indeed, as far as the role of media attention in combination with "free" market forces is concerned, the grim fact is that the more endangered the species to which the particular individual – say, rhino – belongs, the higher its market value: the very last exemplar is the one that will catch the highest price ever, propelling the efforts of professional poachers to have it killed and subsequently sold so as to become a trophy only the very rich can afford and proudly display, testifying in grotesque manner to the perversity of the celebrated "free" interplay of supply and demand as far as the predicament of endangered iconic animals is concerned.

To be sure, the point about human arbitrariness in the face of the suffering of animals is a familiar one, first made when we discussed Midgley. However, the point I want to make here is a different one. In observing how children's perception of wild animals has changed in the course of the last few decades, the novelty worth reflecting upon is to do with what I view as a peculiar sort of sentimentality. Small children feel pity for the big polar bear; they relate to the wild animal not so much in terms of fearing it – "encountering it, I wouldn't stand a chance" – but in terms of worrying about its plight, being struck by the increasingly precarious and stressful life to which it is now doomed, raising questions about whether the last individuals in the wild will be gone during one's lifetime. If ours is an era of extinction for so many wild animals, it is also a time of living with the awareness of extinction on the part of present and future generations of humans.

How to assess this change from being fearful of wild animals owing to their majestic powers, to feeling fearful for their future owing to their appearing so vulnerable in the face of ecological destruction? Just like there is something odd about seagulls roaming garbage cans in urban backyards and polar bears searching for something to eat in Longyearbyen at Svalbard or in Siberian villages, there is something not quite right about children's perception of some of the mightiest mammals on earth changing from the suspense of fear and awe to a pity that borders on sentimentality. Why sentimentality? And why would such a shift in perception be problematic?

If nature's independence *is* its meaning like McKibben famously asserts, perhaps we can say the same about wild animals: their independence is their meaning, their meaning, and their identity, for them, not just for us. Independence would here mean their being what – who – they are meant to be as creatures different from us. For this to be an actuality, the animals have to be free to realize that difference: as Rolston says, wild animals have to be left to their own devices; otherwise their uniqueness will be lost. However, when Rolston stated this in the 1980s, the context was that of conservation, his message that humans should not interfere with the dynamics of life and death as playing out between predator and prey. As a matter of principle, it would never be right to sacrifice, by means of human effort, individuals of one species for the sake of another. Rolston's premise was that, in the long run if not always in the short, nature would take care of itself.

What is historically novel about childrens' perception is that they experience the life of wild animals as *dependent* not independent, spontaneously and as a matter of course. This is not to suggest that animals such as tigers and polar bears are per definition associated with life in captivity in the zoo, where we look at them without their really looking back, so that suspense is replaced by indifference, to recall John Berger. The point is more radical: it is that even if *in situ*, in what we are used to call their original natural habitat, the animals will be seen as living under conditions having less and less to do with being left to their own devices, more and more to do with the impact of our schemes, even – precisely – for *their sake*, the pragmatic argument being that just as pristine wilderness is now a thing of the past, truly independent – autonomous – animals are as well, the task being to keep as many as possible of them in a kind of semi-wild, not yet fully domesticated environment, that is to say, the kind of hybridity between *in situ* and *ex situ* we looked at when discussing Braverman's work.

Notwithstanding the merits of Crist's case for retaining wilderness both conceptually and historically, so as to maintain the possibility of critically exposing the degradation that is occurring, there is a shift from the old normal to the new normal that requires investigation – a shift on the part of human perception of the natural world that reflects profound changes within it, hence a shift that appears reality-oriented and in that sense appropriate, even required, like it or not. Inviting philosophical reflection, the shift is in fact utterly concrete: pictures of albatrosses these days will most likely be pictures of dead birds whose stomachs are filled with plastic; *ditto* pictures of whales, only the plastic will be much bigger in size, just like the animal killed from digesting it, having either mistaken it for food or having been short of something else to eat.

If this is the new normal, what was the old one? Crist reminds us of Aldo Leopold's favourite example of a "biological storm": the passenger pigeons "whose dung would coat thousands of acres, flight would block the sun for hours, and forest roosting extravaganzas would bring mighty branches crashing down" (2019: 123). "Yearly the feathered tempest roared up, down, and across the continent", wrote Leopold, "sucking up the laden fruits of forest and prairie, burning them in a traveling blast of life" (1949: 111). Today, the pigeons are gone, so that "worm and weevil must now perform slowly and silently the biological task that once drew thunder from the firmament". Even so, we are told by "economic moralists" that to mourn the pigeon is mere nostalgia; that "if the pigeoners had not done away with him, the farmers would ultimately have been obliged, in self-defense, to do so". Leopold comments that "this is one of those truths that are valid, but not for the reasons alleged" (ibid.). What reasons? Well, the usual ones in the name of realism, of progress, as were the two somehow identical, two sides of the same coin.

But is there not a fair chance that spreading the news about animals on the brink of extinction will help alert people, perhaps children in particular, to the urgency of the crisis, mobilizing them to act and to put forward political demands, as did the millions who took to the streets worldwide in 2019, following the lead of Greta Thunberg?

For the Levinas-inspired reasons alluded to, the importance of perceiving endangered animals as individuals is not to be underestimated: the face is what commands attention and sparks concern in the case of animals no less than in that of humans. And in both cases the ethical-political potential lies in the capacity of the one highlighted individual to direct attention to the plight of the many. However, to compare the two cases is also to acknowledge the differences between them. If Cecil does the trick for the lions, who does for the vultures? It is not only that some animals appear to us humans as more face-like than others. It is also that some strike us as incomparably more attractive, or precious, or worthy of compassion, than others, this being so for entire species. In other words, there is no match between what is ecologically required and who we are inclined to prioritize based on how we feel about the animals in question. If anything, arbitrariness seems to be an even bigger problem with regard to animals than to humans.

There is a more subtle dimension to the limits of the comparison, however – namely the shift from animals living independent lives to their enduring ever-more dependent such, a shift we first encountered when discussing Paul Shepard's work. It is tempting to suggest that this shift from independence to dependence has no equivalent in human history, lest one considers that Cortez' conquest of America, to allude to Tzvetan Todorov's study (1984), is a case in point, reminding us of how indigenous peoples have been not only persecuted, often massacred, but also, as a form of spiritual extinction going hand in hand with the physical one, have been deprived of their distinctive cosmologies and the various practices that go with them, showing up how the loss of *biodiversity* and the loss of *cultural diversity* are intertwined in today's world, being inseparable aspects of the same ongoing homogenization under the banner of globalization. That said, the shift I am talking about as involving animals can be said to represent a case of "the new normal" in its own right, illustrated when the "average" lion or wolf in the wild is doomed to an existence of quickly diminishing wildness, having its "natural" existence dictated by the "culture" enacted by humans only, to employ Rolston's distinction. It is being monitored all around the clock, purportedly for its own sake, for the greater cause of its preservation in what used to be its natural habitat, comes with the price of tying its every move, its every action, ever closer to the nexus of human-initiated structures and arrangements, risking that the very "wildness" at issue is spoiled instead of safeguarded, making the scare quotes appropriate.

The new normal that emerges from the shift under scrutiny is not only that wild animals are becoming few and far between, or that their wildness qualities are diminishing, a shift whose enormity we can only truly appreciate – measure – by holding onto wilderness the way Crist rightly advocates. What is also new in a sense that commands attention is the willingness on the part of so many of us to adapt to the shift as were it perfectly normal. There is a whole series of concepts conceived to address this adaptability: "shifting baselines syndrome", "generational ecological amnesia", and "extinction of experience" being the most well known. Crist speaks about "the colossal public ignorance surrounding biodiversity's unraveling", the "ongoing, cumulative forgetting of the biosphere's autochtonous nature", and how "dimming knowledge and shriveling experiential horizons surrounding the wealth of planet Earth reveal how the human mind is afflicted by life's destruction" (2019: 14). Ecopsychologist David Kidner makes the point in a way that articulates why retaining the concept of wilderness is important for psychological and symbolic reasons as well as for the empirical ones to do with the critical function of an original blueprint: "Not only is wild nature being destroyed in physical reality, but its obliteration is completed by its elimination from history and imagination, undermining our sense of ourselves as embodied creatures with a natural past, so that a felt resonance with the natural order gives way to a cognitive assimilation into the industrial order" (Kidner in Crist 2019: 115; see Lertzman 2015).

For all its enormity, then, extinction is a phenomenon many people are quite capable of adapting to as perfectly normal. How is that possible, and what does it tell us about our relationship to the animals affected?

VII

To explore these questions, I turn now to the issue of grievability.

Since human-caused extinction of animals amounts to a superkilling of which we are, at least in principle, fully aware, how do we grieve the lives being lost?

The answer is that for the most part, we don't. It is not only that some lives are considered more grievable than others, applying as this (still) does to certain groups of human beings – witness Black Lives Matter. In *Precarious Life: The Powers of Mourning and Violence*, Judith Butler speaks about the "differential allocation of grievability that decides what kind of subject is and must be grieved, and which kind of subject must not, operates to produce and maintain certain exclusionary conceptions of who is normatively human: what counts as a livable life and a grievable death?" (2004: xv). Butler discusses the vulnerability to suffering, to violence and death that we all live with, "a vulnerability to the other that is part of bodily life", a vulnerability that we cannot will away, that is given to all of us as an ineluctable condition of our existence, as are our dependence and our mortality (see Vetlesen 2009). The context for Butler's discussion is political, prompting her to observe that "there are no obituaries for the war casualties that the United States inflicts, and there cannot be". Why? Because if there were to be an obituary, "there would have had to have been a life, a life worth noting, a life worth valuing and preserving, a life that qualifies for recognition" (2004: 34). She offers examples like that of the Palestinian U.S. citizen who submitted to the *San Francisco Chronicle* obituaries for two Palestinian families who had been killed by Israeli troops, only to have their publication rejected because the newspaper "did not want to offend anyone". Butler comments by raising questions like "What might be 'offensive' about the public avowal of sorrow and loss such that obituaries and memorials would function as offensive speech?" and "What is the relation between the violence by which these ungrievable lives were lost and the prohibition on their public grievability?" (2004: 35). Part of her answer is that "there is less a dehumanizing discourse at work here than a refusal of discourse that produces dehumanization as a result" (ibid.).

Following Butler, the crucial task is to establish an acceptance of vulnerability, meant to counter attempts to repudiate it or deny it or to insist that it is tantamount to weakness, a kind of weakness found in certain others and as such, as it were, a reason for attacking them. To truly recognize vulnerability as a *shared* condition is to humanize its bearers, the opposite of using it as a weapon against them; it is to be hopeful that "recognition wields the power to reconstitute vulnerability", actualizing it as a source of concern, of reaching out, not a liability (2004: 43). Butler further develops her position in *The Force of Nonviolence*, focusing, among various groups, on women whose status is that of being "killable", terrorized as they are by the prevalence and impunity of killing practices (2020: 189).

It will not be lost on the reader that Butler concentrates on humans and humans only in her account of lives deemed ungrievable and individuals (or groups) regarded as killable. The anthropocentrism of her approach may come as a surprise, or even as a disappointment in being so conventionally held on to by a prominent radical thinker, raising questions about how deliberate it is and whether Butler could have included other-than-humans without undermining her objectives. I shall not attempt to answer these questions here. The relevance of her main points and observations should be evident even when what we want to explore is not the grievability or rather denial of such of (certain) humans but of animals, be it some of them or all of them.

Common to both cases is a denial of vulnerability as encountered in the other. Granted. But that doesn't take us very far. The sort of denial crucially involved here is the denial of the worthiness of the bearer of vulnerability. This is what is involved in the recognition Butler talks about: that vulnerability is *a basis for being worthy of recognition*, not a feature tantamount to being weak or inferior and so as deficient in moral standing. The morally important thing about vulnerability, then, is that it calls for a certain response – the response of responsibility.

Of course, this is the view we found promoted in Levinas' ethics, the problem being however that he is commonly taken to reserve the ethical significance of the connection between vulnerability and responsibility to humans. But what about animals? How to access, how to recognize *their* vulnerability? And how to do so in a way that sheds light on the peculiar nature of *their* grievability?

In making an analytic shift from humans to animals, Hegelian accounts of recognition and Levinasian accounts of responsibility come up against a limit: quite simply, they are not very helpful. Suffice it to say that they are too general and abstract to do justice to the historical and ecological circumstances in which animals are now threatened with extinction.

The losses that add up to the extinction of a species are utterly concrete: so many occurrences in a particular time and place affecting individual animals. As Deborah Bird Rose reminds us, life processes take place in time and are emplaced. This also holds for loss. Whereas scientific analyses of environmental crisis typically deal in agglomerated figures and statistics that are largely abstracted from place, loss is primarily experienced *in place*. "We are", writes Rose, "unlikely to experience the predicted loss of twenty-one out of twenty-four species of butterflies in Australia, but we are almost certain to encounter absences – places where butterflies used to be and are no more", be they forests or paddocks or one's own backyard (2004: 49).

To illustrate, consider penguins, penguins in a particular place such as the thin brand of rocky foreshore at Manly, inside the mouth of Sydney Harbour, Australia. Here nests a tiny colony of Little Penguins. What is their situation like? In recent years, Thom van Dooren tells us, an estimated 50 percent of the shoreline in question, hundreds of miles in length, have seen the building of seawalls and other built structures; the homeowners who can afford it erect seawalls in order to moor a boat or build a swimming pool. Each year, the shoreline becomes more built up, more noisy, more dangerous for the penguins who come there to breed. "And yet", writes van Dooren, "year after year, the penguins keep returning". This is the fact, this is the image that continues to haunt him: a penguin returning to a burrow, to a breeding place, that is no longer there or has been "transformed so dramatically that it is no longer habitable" (2014: 66). So, while the people who buy the houses and build the seawalls are attracted to the place as a chance to be "closer to nature", the cost paid by the penguins comes in the form of their returning to lost places.

This is the bigger picture of which the penguins in Sydney Harbour is but one of numerous examples: "All over the world, animals are drawn to return faithfully to places that no longer exist", van Dooren observes (ibid.). The site fidelity of so many species of animals testifies to their being *philopatric*, a term that literally means "love of one's home", employed in biology to describe a process in which an animal returns to its place of birth or hatching to reproduce; such loyalty to place is crucial in enabling birds to retain high-quality nests over many years, spanning generations. Alongside inheriting genes, organisms – animals – also inherit environments: not any piece of land or shoreline will do, only one is "home"; more than the sum of their ecological parts, the one place that is home carries the histories and the stories of the particular animals who belong there.

To use a term introduced earlier but not employed by van Dooren, *silent killing* is the *modus operandi* of the dynamics that cause more and more species of animals to go extinct. Technically speaking and in bureaucratic terminology, "alteration in the use of land" is the main cause: the motorways, housing complexes, shopping malls, parking lots, factories, mines, livestock grasslands, palm oil plantations, and so on and so forth, here and there and everywhere, put up in the name of progress and development all over the world, in each and every instance taking over, converting for novel use and exploitation what remains of "untapped natural resources", such as the shoreline that used to be the little penguins home, the forest that used to be the wolves' home, the rivers and lakes that used to be the beavers' home, and so on, the homes for animals everywhere on Earth.

Silent killing kills by stealing the homes of animals, snatching them away from them one place after the other, year after year, day after day. The causal sequence is neither subtle nor particularly complex. To the contrary, it is as straightforward as it gets, that of zero-sum: each new place converted from an "undeveloped" to a "developed" state is a place taken over for human use *by way of* being lost as a home for other-than-humans of all sorts, animals in particular.

Despite being utterly concrete and real for the animals affected, the process is perfectly abstract and invisible for most of us humans. True, we may encounter absences like Rose mentions, say, places where butterflies used to be and are no more. But noticing absences is only possible if you are sufficiently familiar with a particular place to actually make the comparison and thereby spot the difference between what that particular place used to be like and what it has now become – precisely the exercise of experience-based comparison between past and present that is undermined by the psychological mechanisms referred to above, "shifting baseline syndrome" in particular. The upshot is that "with each ensuing generation, the amount of environmental degradation increases, but each generation in its youth takes that degraded condition as the nondegraded condition – as the normal experience" (Peterson 2009: 92).

"Eco-grief" and "ecological mourning" are conceptual innovations aimed to address the emotional response that the loss of animals, habitats, and sometimes entire landscapes – the landscape of childhood – may elicit (see Griffiths 2013). The Australian ecophilosopher Glenn Albrecht has coined the term "solastalgia", defining it as "the pain and distress caused by the ongoing loss of solace and the sense of desolation connected to the present state of one's home and territory. It is the existential and lived experience of negative environmental change, manifest as an attack on one's sense of place". Solastalgia is the homesickness you experience when at home, born of the memory of what home used to be like, home as shared and co-constituted by so many "others", animals in particular (2019: 38).

What complicates mourning in the face of ongoing loss and extinction is the fact that we are complicit in their occurrence. The fact of complicity – though vastly, even grotesquely, unevenly distributed among nations, regions, and socio-economic classes – makes the experience of grief at extinction unwelcome, its articulation and public display problematic. As a culture and as a society we lack traditions and rituals for collective mourning of animals or landscapes lost, in effect privatizing the psychic toll of the loss and the emotional reactions to it, and so depoliticizing it. As citizens of present-day Western societies we are not expected to feel, let alone openly display, grief over the loss of animals (except one's pets), be it as particular individuals or as species, excluding animals in this regard just like in their appropriateness as "significant others", discussed earlier. Indeed, it's not only that we as a collective lack practices of commemoration like burial ceremonies and newspaper obituaries to acknowledge that the animals are now gone one species after the other, from one place after the other, the pace of loss speeding up rather than slowing down year after year. We lack these sort of practices because as a culture we persist in denying that non-humans such as animals are worthy of being grieved in the first place – they are denied grievability in Butler's sense. As long as they continue to be so, the culturally entrenched deficit just noted will continue to raise its ugly head in the form of attacking those who publicly display their grief, the more so if such individuals, Greta Thunberg-like, dare to point to the human, societal, and political culpability for the loss, pointing their finger at powerful agents particularly responsible for it, such as the oil company Equinor in my native Norway.

In fact, the aggression with which acts of public display of eco-grief and eco-mourning are being met, overwhelmingly by men, demonstrates that to speak about a "lack" of cultural practices (rituals), though correct, risks putting it much too mildly: the intolerance and outright anger with which these emotions are met show that expressing them publicly amounts to breaking a cultural taboo. Indeed, there is but a short way from privatizing and depoliticizing these emotions to downright pathologizing them, stigmatizing as "mental illness" what should be regarded as "rational responses to irrational situations" (Albrecht 2017: 311; see Nicholson 2003: 43). The culturally sustained sense of entitlement, of the right to have it all, to choose individually and freely, without interference from the authorities or "moralizing" from other citizens, what to do, what to buy, where to travel, as trumping all other considerations and the rights and freedoms of everybody else, is clearly what is offended once individuals like Thunberg points her finger at the societal roots of the ongoing devastation of nature, roots better left unexamined and unidentified, lest the grand narrative about progress be challenged. If anything, there is *too little* mourning, understood as a perfectly justified response to an intolerable situation, making its public articulation into an act of resistance, of non-compliance in the proud tradition of non-violent protest that Butler rightly champions.

The upshot is that, culturally as well of psychologically, the framework for mourning must be expanded so as to accommodate mourning for other-than-humans as a perfectly sane, non-irrational and non-pathological variant of mourning. Morally and politically, to acknowledge such mourning as perfectly appropriate to its object, even called for, is to recognize these losses as worthy of being mourned, passing the burden of having to explain themselves from those who feel this way to those who don't.

This raises questions about responsibility and guilt for which not only mainstream political and moral theory is ill-prepared, as exemplified by Habermas and Rawls (see Mulgan 2011); so is psychology as well. Take Freud's seminal paper "Mourning and Melancholia" from 1917. Freud takes it for granted that the kind of mourning worth talking about and in need of treatment is that manifest in one person's experience at the loss of another. All parties to the loss involved, to the deprivation suffered, are human persons as a matter of course, the mourned lost object as well as the mourning subject. To the degree that the natural world at all enters into the picture, it does so in the form of a piece of reality taken to be intact and solid in a situation otherwise shot through with the disturbance incurred by the loss suffered. Nature – typically in the form of a particular place, often cherished since childhood – is relevant to the process (Freud: work) of mourning in being attributed a role in helping keep things stable when emotionally and psychically they are not. Nature, that is to say, is trusted as suited for purposes of holding and containment, to use Melanie Klein's and Donald Winnicott's psychoanalytic terms, nature taken as being there "for me", always, unfailingly and loyally, whenever my bond with other humans is cut short, revealing the *precariousness* of inter-human relationships that so starkly contrasts with the *permanence* and calm stability of nature, being always itself, undisturbed by whatever disturbs us humans. Hence nature will allow me to regain my vitality, to renew my interest in the possibilities that lie ahead, in short, to prioritize the future over the past and so helping me let go of the things I cannot undo, psychically liberating me from a painful loss in the past so as to thrust myself forward into a future holding the promise of new loves and the forging of new bonds.

Evidently, that one day the natural world would lose these seemingly rock solid characteristics, and that it would do so by way of degradation and destruction inflicted by us humans, is a possibility not entertained by Freud, understandably preoccupied at the time with the devastation of the first world war. A century later, nature is subject to such devastation as to alter it substantially – not only in the sense of the emerging Earth science of the Anthropocene, but also in a sense crucial to psychology in general and perhaps psychoanalysis in particular. Instead of nature being there "for us", meeting our need for comfort in times of trouble, we now confront a nature so damaged as to fail us in these expectations – failing us because we have failed it. That nature itself would become the object to be mourned is what is truly new.

Implied in this shift of focus from the inter-human to the natural world is also a shift in temporality, more subtle at that and so far receiving far less attention. Freud, here at one with common sense, took it for granted that the cause prompting the hard work of mourning would be a thing – an event, an object – of the past, so that as time goes by, the more distant will become the source and the point of origin of the entire process, a notion attested to in sayings like "trust that time will help heal the wound". In contrast to being past-oriented like this, the sort of mourning we are only now starting to grasp, because only now starting to experience and to have objective grounds for experiencing, is eminently *future*-oriented: the danger we are alerted to, the trends we find alarming, are not things of the past whose impact peters out as time passes. To the contrary, the damage and decimation to the point of extinction of species, together with the disappearance of entire habitats and biomes, are part of a trajectory of destruction such that we "ain't seen nothing yet", such that, expectedly, things will get worse, there therefore being good reasons to fear for the future of all involved. This means that the joy we today may feel at landscapes not altered, animals not lost, gives comfort only up to a point: in so many cases, the *not* is a *not yet*, knowing as we do that so much is *in the process* of being lost.

To conclude, what we need to recognize is not only that we may mourn animals too, not only fellow humans; and that if we haven't mourned animals as part of our cultural tradition, we should learn to do so now, however complicated that learning may be owing to our complicity in their plight. Last but not least, animals are themselves no strangers to mourning, as everyone who has heard Jane Goodall address the phenomenon will know. That animals have more reason to mourn for each other now than ever before speaks to the magnitude of the loss we are inflicting upon them.

Bibliography

Adorno, Theodor W., 1973 [1966]. *Negative Dialectics*. New York, NY: Seabury Press.
———, 2000. *Problems of Moral Philosophy*. Oxford: Polity Press.
———, 2005. *Critical Models*. New York, NY: Columbia University Press.
Albrecht, Glenn, 2017. "Solastalgia and the New Mourning", in Cunsolo, Ashlee and Karen Landman, eds., 292–315.
———, 2019. *Earth Emotions: New Words for a New World*. Ithaca, NY: Cornell University Press.
Alford, C. Fred, 1992. *The Psychoanalytic Theory of Greek Tragedy*. New Haven, CT: Yale University Press.
———, 2002. *Levinas, the Frankfurt School and Psychoanalysis*. New York, NY: Continuum.
Angus, Ian, 2016. *Facing the Anthropocene: Fossil Capitalism and the Crisis of Earth System*. New York, NY: Monthly Review Press.
Arendt, Hannah, 1951. *The Origins of Totalitarianism*. New York, NY: Harcourt.
———, 1958. *The Human Condition*. Chicago, IL: The University of Chicago Press.
———, 1965. *Eichmann in Jerusalem. A Report on the Banality of Evil*. New York, NY: Viking Press.
———, 1978. *The Life of the Mind*. London: Secker & Warburg.
Baier, Annette, 1991. *A Progress of Sentiments*. Harvard: Harvard University Press.
Barad, Karen, 2007. *Meeting the Universe Halfway*. Durham: Duke University Press.
Barnosky, Anthony D., 2009. *Heatstroke*. Washington, DC: Island Press.
Bastian, Michelle, 2017. "Encountering Leatherbacks in Multispecies Knots of Time", in Rose, Deborah Bird et al., eds., 149–186.
Bauman, Zygmunt, 1989. *Modernity and the Holocaust*. Oxford: Polity Press.
———, 1993. *Postmodern Ethics*. Oxford: Blackwell.
Beauvoir, Simone de, 1952 [1949]. *The Second Sex*. New York, NY: Harper & Row.
Beck, Ulrich, 1986. *Risk Society: Towards a New Modernity*. London: Sage.
Bekoff, Marc, and Dale Jamieson, eds., 1990. *Interpretation and Explanation in the Study of Animal Behavior*. Boulder, CO: Westview Press.
Bentham, Jeremy, 1948 [1789]. *The Principles of Morals and Legislation*. New York, NY: Hafner.
Benton, Tim, 2001. "Why Are Sociologists Naturephobes?", in Lopez, Jennifer and G. Potter, eds., 132–145.
Berger, John, 1980. *About Looking*. New York, NY: Pantheon.

Berkes, Fikret, 2008. *Sacred Ecology*. London: Routledge.
Bernasconi, Robert, and David Wood, eds., 1988. *The Provocation of Levinas*. London: Routledge.
Bernasconi, Robert, and Simon Critchley, eds., 1991. *Re-Reading Levinas*. London: Athlone.
Bernstein, Jay M., 2001. *Adorno: Disenchantment and Ethics*. Cambridge: Cambridge University Press.
Bettelheim, Bruno, 1976. *The Uses of Enchantment*. Harmondsworth: Penguin.
Boltanski, Luc, 1999. *Distant Suffering*. Cambridge: Cambridge University Press.
Braverman, Irus, 2015. *Wild Life. The Institution of Nature*. Stanford, CA: Stanford University Press.
Buber, Martin, 1970. *I and Thou*. New York, NY: Scribner.
Butler, Judith, 2004. *Precarious Life: The Powers of Mourning and Violence*. London: Verso.
_____, 2020. *The Force of Non-Violence. An Ethico-Political Bind*. London: Verso.
Callicott, J. Baird, 1989. *In Defense of the Land Ethic*. New York, NY: SUNY.
_____, 1994. *Earth's Insights*. Berkeley, CA: University of California Press.
Carson, Rachel, 1962. *Silent Spring*. Boston, MA: Houghton Mifflin.
Cavalieri, Paola, 2009. *The Death of the Animal*. New York, NY: Columbia University Press.
Cavell, Stanley et al., 2008. *Philosophy and Animal Life*. New York, NY: Columbia University Press.
Ceballos, Gerardo, Paul Ehrlich, and Rodolfo Dirzo, 2017. "Biological annihilation via the ongoing sixth mass extinction signaled by vertebrate population losses and declines", in *Proceedings of the National Academy of Sciences of the United States of America* 114 (30): E6089–E6096.
Chagnon, Napoleon, 1998. *Yanomamo*. New York, NY: Harcourt Brace.
Challenger, Melanie, 2021. *How to Be Animal. A New History of What It Means to Be Human*. Edinburg: Canongate.
Cobb, Edith, 1977. *The Ecology of Imagination in Childhood*. Oxford: Oxford University Press.
Coetzee, John, 1997. *Boyhood: Scenes from Provincial Life*. London: Secker and Warburg.
_____, 1999a. *Disgrace*. London: Secker and Warburg.
_____, 1999b. *The Lives of Animals*. Princeton, NJ: Princeton University Press.
_____, 2003. *Elizabeth Costello*. London: Secker and Warburg.
Cook, Deborah, 2011. *Adorno on Nature*. Durham: Acumen.
Crary, Alice, 2015. *Inside Ethics. On the Demands of Moral Thought*. Harvard: Harvard University Press.
Crist, Eileen, 1999. *Images of Animals: Anthropomorphism and Animal Mind*. Philadelphia, PA: Temple University Press.
_____, 2019. *Abundant Earth: Toward an Ecological Civilization*. Chicago, IL: University of Chicago Press.
Critchley, Simon, 1992. *The Ethics of Deconstruction: Derrida and Levinas*. Oxford: Blackwell.
_____, and Robert Bernasconi, eds., 2002. *The Cambridge Companion to Levinas*. Cambridge: Cambridge University Press.
Cunsolo, Ashlee, and Karen Landman, eds., 2017. *Mourning Nature: Hope at the Heart of Ecological Loss and Grief*. Montreal: McGill-Queen's University Press

Darwin, Charles, 1965 [1872]. *The Expression of Emotions in Man and Animals.* Chicago, IL: University of Chicago Press.
_____, 1981 [1871]. *The Descent of Man.* Princeton, NJ: Princeton University Press.
_____, 1985 [1859]. *The Origin of Species.* Harmondsworth: Penguin.
Derrida, Jacques, 1999. *Adieu to Emmanuel Levinas.* Stanford, CA: Stanford University Press.
_____, 2008. *The Animal That Therefore I Am.* New York, NY: Fordham University Press.
Descartes, René, 1978a [1637]. *A Discourse on Method.* London: Dent.
_____, 1978b [1644]. *Principles of Philosophy.* London: Dent.
_____, 1981 [1629–1649]. *Philosophical Letters.* Minneapolis, MN: University of Minnesota Press.
_____, 1985 [1641]. *Meditations.* Cambridge: Cambridge University Press.
Descola, Philippe, 2013. *Beyond Nature and Culture.* Chicago, IL: The University of Chicago Press.
Despret, Vinciane, 2022. "Responding Bodies and Partial Affinities in Human-Animal Worlds", in Kalof, Linda and Amy Fitzgerald, eds., op. cit.: 539–559.
Dodds, Joseph, 2011. *Psychoanalysis and Ecology at the Edge of Chaos.* London: Routledge.
Evernden, Neil, 1985. *The Natural Alien.* Toronto: University of Toronto Press.
_____, 1993. *The Social Creation of Nature.* Baltimore, MD: The Johns Hopkins University Press.
Fisher, Andy, 2013. *Radical Ecopsychology.* New York, NY: SUNY.
Foltz, Bruce, and Robert Frodeman, eds., 2004. *Rethinking Nature.* Bloomington, IL: Indiana University Press.
Freud, Sigmund, 1984. *On Metapsychology.* Harmondsworth: Penguin.
_____, 1985 [1930]. *Civilization and Its Discontents.* Harmondsworth: Penguin.
Fromm, Erich, 1955. *The Sane Society.* London: Routledge.
Gadamer, Hans-Georg, 1975 [1960]. *Truth and Method.* London: Bloomsbury.
Giddens, Anthony, 1984. *The Constitution of Society.* Oxford: Polity Press.
Gilligan, Carol, 1982. *In a Different Voice.* Harvard: Harvard University Press.
Goodall, Jane, 1971. *In the Shadow of Man.* New York, NY: Collins.
Goodpaster, Kenneth E., ed., 1979. *Ethics and Problems of the 21st Century.* Notre Dame, IN: University of Notre Dame Press.
Gordon, Peter E. et al., eds., 2018. *The Routledge Companion to the Frankfurt School.* New York, NY: Routledge.
Griffiths, Jay, 2013. *Kith: The Riddle of the Childscape.* London: Hamilton.
Grusovnik, Tomaz et al., eds., 2021. *Environmental and Animal Abuse Denial.* London: Lexington Books.
The Guardian Weekly, 2021. "Bird Species at Risk of Extinction", January 15, 2021: 31.
Habermas, Jürgen, 1971. *Toward a Rational Society.* London: Heinemann.
_____, 1972 [1968]. *Knowledge and Human Interests.* London: Heinemann.
_____, 1979. *Communication and the Evolution of Society.* London: Heinemann.
_____, 1982. "A Reply to My Critics", in Thompson, John and David Held, eds., 219–283.
_____, 1984/1987 [1981]. *The Theory of Communicative Action, I & II.* Boston, MA: Beacon Press.
_____, 1990. *Moral Consciousness and Communicative Action.* Cambridge: Polity Press.

_____, 1993. *Justification and Application*. Cambridge, MA: MIT Press.

_____, 2012. *Nachmetaphysisches Denken*. Berlin: Suhrkamp.

_____, 2019. *Auch eine Geschichte der Philosophie*. Berlin: Suhrkamp.

Hacking, Ian, 2008. "Deflections", in Stanley, Cavell et al., eds., 139–172.

Hamilton, Clive, 2017. *Defiant Earth: The Fate of Humans in the Anthropocene*. Oxford: Polity Press.

Haraway, Donna J., 2008. *When Species Meet*. Minneapolis, MN: University of Minnesota Press.

Hatley, James, 2004. "The Uncanny Goodness of Being Edible to Bears", in Foltz, Bruce and Robert Frodeman, eds., 13–31.

Hegel, Georg W.F., 1977 [1807]. *Phenomenology of Spirit*. Oxford: Oxford University Press.

_____, 1991 [1821]. *Elements of the Philosophy of Right*. Cambridge: Cambridge University Press.

Heidegger, Martin 1962 [1927]. *Being and Time*. New York, NY: Harper & Row.

_____, 1982. *The Basic Problems of Phenomenology*. Bloomington, IN: Indiana University Press.

Heise, Ursula, 2016. *Imagining Extinction: The Cultural Meanings of Endangered Species*. Chicago, IL: The University of Chicago Press.

Hilberg, Raul, 1985. *The Destruction of the European Jews*. New York, NY: Holmes and Meier.

Hobbes, Thomas, 1966a [1651]. *Leviathan*, in *English Works, Vol. 3*. Aalen: Scientia.

_____, 1966b [1655]. *De Corpore*, in *English Works, Vol. 11*. Aalen: Scientia.

Horkheimer, Max, and Theodor W. Adorno, 1979 [1947]. *Dialectic of Enlightenment*. London: Verso.

Hornborg, Alf, 2019. *Nature, Society, and Justice in the Anthropocene*. Cambridge: Cambridge U.P.

Huhn, Tom, 2004. *The Cambridge Companion to Adorno*. Cambridge: Cambridge University Press.

Hume, David, 1969 [1740]. *A Treatise of Human Nature*. Harmondsworth: Penguin.

Ingold, Tim, 2011. *The Perception of the Environment*. London: Routledge.

Irigaray, Luce, 1991. "Questions to Emmanuel Levinas: On the Divinity of Love", in Bernasconi, Robert and Simon Critchley, eds., 109–118.

Jonas, Hans, 1984 [1979]. *The Imperative of Responsibility*. Chicago, IL: University of Chicago Press.

_____, 2001 [1966]. *The Phenomenon of Life: Toward a Philosophical Biology*. Evanston: Northwestern University Press.

Jones, Lucy, 2020. *Losing Eden. Why Our Minds Need the Wild*. London: Allen Lane.

Kalof, Linda, and Amy Fitzgerald, eds., 2022. *The Animals Reader*. Second Edition. New York, NY: Routledge.

Kant, Immanuel, 1951 [1790]. *Critique of Judgment*. New York, NY: Hafner Press.

_____, 1930. *Lectures on Ethics*. New York, NY: Harper & Row.

Kessler, Neil, 2019. *Ontology and Closeness in Human-Nature Relationships*. Cham: Springer.

Klein, Melanie, 1988. *Love, Guilt and Reparation*. London: Virago.

_____, 1989. *Envy and Gratitude*. London: Virago.

Klein, Naomi, 2014. *This Changes Everything. Capitalism vs the Climate*. London: Allen Lane.

Kohn, Eduardo, 2013. *How Forests Think: Toward an Anthropology Beyond the Human*. Berkeley, CA: University of California Press.
Kolbert, Elizabeth, 2014. *The Sixth Extinction. An Unnatural History*. London: Bloomsbury.
Kopenawa, Davi, and Bruce Albert, 2013. *The Falling Sky. Words of a Yanomami Shaman*. Cambridge, MA: Belknap Press.
Latour, Bruno, 2004. *Politics of Nature*. Cambridge, MA: Harvard University Press.
——, 2017. *Facing Gaia. Eight Lectures on the New Climate Regime*. Oxford: Polity Press.
Lear, Jonathan, 2006. *Radical Hope: Ethics in the Face of Cultural Devastation*. Cambridge, MA: Harvard University Press.
Lee Mueller, Martin, 2017. *Being Salmon, Being Human*. Vermont: Chelsea Green.
Leopold, Aldo, 1949. *A Sand County Almanac*. Oxford: Oxford University Press.
Lertzman, Renee, 2015. *Environmental Melancholia: Psychoanalytic Dimensions of Engagement*. New York, NY: Routledge.
Levinas, Emmanuel, 1985. *Ethics and Infinity*. Pittsburgh, PA: Duquesne Press.
——, 1987. *Time and the Other*. Pittsburgh, PA: Duquesne Press.
——, 1988a. "Useless Suffering", in Bernasconi, Robert and David Wood, eds., 156–167.
——, 1988b. "The Paradox of Morality", in Bernasconi, Robert and David Wood, eds., 168–180.
——, 1990. "The Name of a Dog", in *Difficult Freedom*. Baltimore, MD: Johns Hopkins University Press, 151–155.
——, 1991a. *Totality and Infinity*. Dordrecht: Kluwer.
——, 1991b. *Otherwise than Being or Beyond Essence*. Dordrecht: Kluwer.
——, 1993a. *Collected Philosophical Papers*. Dordrecht: Kluwer.
——, 1993b. *Outside the Subject*. London: Athlone Press.
Lloyd, Genevieve, 1993. *The Man of Reason: "Male" and "Female" in Western Philosophy*. London: Routledge.
Lopez, Jennifer, and G. Potter, eds., 2001. *After Postmodernism: An Introduction to Critical Realism*. London: Athlone Press.
Lorenz, Konrad, 1966. *On Aggression*. London: Methuen.
——, 1976. *Behind the Mirror*. London: Methuen.
Louv, Richard, 2005. *Last Child in the Woods. Saving Our Children from Nature-Deficit Disorder*. London: Atlantic Books.
Malm, Andreas, 2018. *The Progress of this Storm*. London: Verso.
Marcuse, Herbert, 1956. *Eros and Civilization: A Philosophical Inquiry into Freud*. London: Routledge.
Margulis, Lynn, and Dorion Sagan, 2000. *What Is Life?* Berkeley, CA: University of California Press.
Marris, Emma, 2021. *Wild Souls. Freedom and Flourishing in the Non-Human World*. New York, NY: Bloomsbury.
Mathews, Freya, 1991. *The Ecological Self*. London: Routledge.
——, 2016. *Without Animals Life Is Not Worth Living*. Port Adelaide: Ginninderra Press.
McKibben, Bill, 1989. *The End of Nature: Humanity, Climate Change, and the Natural World*. New York, NY: Random House.
Menke, Christoph, 2004. "Genealogy and Critique", in Huhn, Tom ed., 302–326.
Merchant, Caroline, 1990 [1980]. *The Death of Nature*. New York, NY: Harper.

———, 2005. *Radical Ecology. The Search for a Livable World.* New York, NY: Routledge.
———, 2016. *Autonomous Nature.* New York, NY: Routledge.
Midgley, David, ed., 2005. *The Essential Mary Midgley.* London: Routledge.
Midgley, Mary, 1995 [1978]. *Beast and Man. The Roots of Human Nature.* London: Routledge.
———, 1998 [1983]. *Animals and Why They Matter.* Athens, GA: University of Georgia Press.
———, 2001. *Science and Poetry.* London: Routledge.
———, 2004. *The Myths We Live by.* London: Routledge.
———, 2010. *The Solitary Self. Darwin and the Selfish Gene.* Durham: Acumen.
———, 2014. *Are You an Illusion?* Durham: Acumen.
———, 2018. *What is Philosophy For?* New York: Bloomsbury.
Monbiot, George, 2013. *Feral. Searching for Enchantment on the Frontiers of Rewilding.* London: Allen Lane.
Morton, Timothy, 2013. *Hyperobjects: Philosophy and Ecology After the End of the World.* Minneapolis, MN: University of Minnesota Press.
Mulgan, Tim, 2011. *Ethics for a Broken World. Imagining Philosophy After Catastrophe.* New York, NY: Routledge.
Murdoch, Iris, 1970. *The Sovereignty of Good.* London: Routledge.
Nagel, Thomas, 1979. "What Is It like to Be a Bat?", in *Mortal Questions.* Cambridge: Cambridge University Press, 165–180.
Netz, Reviel, 2004. *Barbed Wire: An Ecology of Modernity.* Middletown, CT: Wesleyan University Press.
Neyrat, Frederic, 2019. *The Unconstructable Earth.* New York, NY: Fordham.
Nicholson, Shierry Weber, 2003. *The Love of Nature and the End of the World.* Cambridge, MA: MIT Press.
Næss, Arne, 1989. *Ecology, Community, and Lifestyle.* Cambridge: Cambridge U.P.
Orange, Donna, 2017. *Climate Crisis, Psychoanalysis, and Radical Ethics.* New York, NY: Routledge.
Ottmann, Henning, 1982. "Cognitive Interests and Self-Reflection", in Thompson, John and David Held, eds., 79–97.
Peterson, Anna L., 2009. *Everyday Ethics and Social Change.* New York, NY: Columbia University Press.
Platonov, Andrei, 2008. *Soul and Other Stories.* New York, NY: NURB.
Plumwood, Val, 1993. *Feminism and the Mastery of Nature.* London: Routledge.
———, 1996. "Being Prey", in *Terra Nova* 1 (3): 32–44.
———, 2002. *Environmental Culture. The Ecological Crisis of Reason.* New York, NY: Routledge.
———, 2004. "Taken by a Crocodile", in *The Age*, January 12.
———, 2012. *The Eye of the Crocodile.* Melbourne: ANY Press.
Pollan, M., 2006. *The Omnivore's Dilemma: A Natural History of Four Meals.* United Kingdom: Bloomsbury Publishing.
Rawls, John, 1972. *A Theory of Justice.* Oxford: Oxford University Press.
———, 1993. *Political Liberalism.* New York, NY: Columbia University Press.
Regan, Tom, 1983. *The Case for Animal Rights.* Berkeley, CA: University of California Press.
Ricoeur, Paul, 1992. *Oneself as Another.* Chicago, IL: University of Chicago Press.

Rollin, B. E., 1990. "How the Animals Lost Their Minds", in Bekoff, Marc and Dale Jamieson, eds., 375–393.
Rolston, Holmes, 1988. *Environmental Ethics. Duties to and Values in the Natural World*. Philadelphia, PA: Temple University Press.
Rose, Deborah Bird, 2000 [1992]. *Dingo Makes Us Human. Life and Land in an Australian Aboriginal Culture*. Cambridge: Cambridge University Press.
———, 2004. *Reports from a Wild Country. Ethics for Decolonisation*. Sydney: UNSW Press.
———, 2011. *Wild Dog Dreaming. Love and Extinction*. Charlottesville, VA: University of Virginia Press.
———, et al., eds., 2017. *Extinction Studies. Stories of Time, Death, and Generations*. New York, NY: Columbia University Press.
Sahlins, Marshall, 1972. *Stone Age Economics*. London: Tavistock.
Sandler, Roland, 2012. *The Ethics of Extinction*. Cambridge: Cambridge University Press.
Sartre, Jean-Paul, 1956 [1943]. *Being and Nothingness*. New York, NY: Washington Square Press.
———, 1958. *Existentialism and Humanism*. London: Methuen.
Searles, Harold, 1960. *The Nonhuman Environment in Normal Development and in Schizophrenia*. New York, NY: International Universities Press.
Shah, Sonia, 2017. *Pandemic*. New York, NY: Picador.
Shepard, Paul, 1982. *Nature and Madness*. Athens, GA: University of Georgia Press.
———, 1997. *The Others. How Animals Made Us Human*. Washington, DC: Island Press.
———, 1998 [1973]. *The Tender Carnivore and the Sacred Game*. Athens, GA: University of Georgia Press.
Singer, Peter, 1976. *Animal Liberation*. London: Jonathan Cape.
———, 1979. "Not for Human Only: The Place of Nonhumans in Environmental Issues", in Goodpaster, K.E., ed., 191–206.
Soper, Kate, 1995. *What Is Nature?* Oxford: Blackwell.
Smuts, Barbara, 2001. "Encounters with Animal Minds", in *Journal of Consciousness Studies* 8, 293–309.
Steiner, Gary, 2008. *Animals and the Moral Community. Mental Life, Moral Status and Kinship*. New York, NY: Columbia University Press.
———, 2013. *Animals and the Limits of Postmodernism*. New York, NY: Columbia University Press.
Stern, Daniel, 1985. *The Interpersonal World of the Infant*. New York, NY: Basic Books.
Sverdrup-Thygeson, Anne, 2021. Personal communication.
Taylor, Charles, 1989. *Sources of the Self*. Cambridge: Cambridge University Press.
Taylor, Paul W., 2011 [1986]. *Respect for Nature. A Theory of Environmental Ethics*. Princeton, NJ: Princeton University Press.
Tester, Keith, 1991. *Animals and Society. The Humanity of Animal Rights*. London: Routledge.
Thompson, John, and David Held, eds., 1982. *Habermas. Critical Debates*. London: Macmillan.
Thunberg, Greta, 2020. *No One Is Too Small to Make a Difference*. New York, NY: Penguin.

Timofeeva, Oxana, 2018. *The History of Animals. A Philosophy.* New York, NY: Bloomsbury.

Tinbergen, Nikolaas, 1951. *The Study of Instinct.* Oxford: Oxford University Press.

_____, 1953. *Social Behavior in Animals.* London: Methuen.

_____, 1961. *The Herring Gull's World.* New York, NY: Basic Books.

Todorov, Tzvetan, 1984. *The Conquest of America. The Question of the Other.* New York, NY: Harper.

Tomasello, Michael, 2012. *A History of Human Thought.* London: Yale University Press.

van Dooren, Thom, 2014. *Flight Ways. Life and Loss at the Edge of Extinction.* New York, NY: Columbia University Press.

Vetlesen, Arne Johan, 1994. *Perception, Empathy, and Judgment. An Inquiry into The Preconditions of Moral Performance.* University Park, PA: Penn State University Press.

_____, and Harald Jodalen, eds., 1997. *Closeness. An Ethics.* Oslo and Boston: Scandinavian University Press.

_____, 2005. *Evil and Human Agency. Understanding Collective Evildoing.* Cambridge: Cambridge University Press.

_____, 2009. *A Philosophy of Pain.* London: Reaktion Books.

_____, 2015. *The Denial of Nature. Environmental Philosophy in the Era of Global Capitalism.* Abingdon and New York: Routledge.

_____, 2017. "Ethics and Value in Næss' Ecophilosophy", in *Worldviews* 21, 251–261.

_____, 2018. "Critical Theory and the Environment", in Gordon, Peter E. et al., eds., 471–485.

_____, 2019. *Cosmologies of the Anthropocene. Panpsychism, Animism, and the Limits of Posthumanism.* Abingdon and New York: Routledge.

_____, 2021. "Denial as a Sense of Entitlement. Assessing the Role of Culture", in Grusovnik, Tomaz et al., eds., 35–54.

Viveiros de Castro, Eduardo, 2015. *The Relative Native. Essays on Indigenous Conceptual Worlds.* Chicago, IL: Hau Books.

Walker, Alice, 1989. *Living by the Word.* San Diego, CA: Harcourt.

Weintrobe, Sally, ed., 2013. *Engaging with Climate Change. Psychoanalytic and Interdisciplinary Perspectives.* London: Routledge.

Wendt, Alexander, 2015. *Quantum Mind and Social Science.* Cambridge: Cambridge University Press.

Whitehead, Alfred North, 1967 [1925]. *Science and the Modern World.* New York, NY: The Free Press.

Wiggins, David, 1987. *Needs, Values, Truth.* Oxford: Blackwell.

Willett, Cynthia, 2014. *Interspecies Ethics.* New York, NY: Columbia University Press.

Williams, Robert, 1997. *Hegel's Ethics of Recognition.* Berkeley, CA: University of California Press.

Wilson, Edward O., 1975. *Sociobiology. The New Synthesis.* Cambridge, MA: Harvard University Press.

Winnicott, Donald W., 2016 [1951]. "Transitional Objects and Transitional Phenomena", in *Collected Works of Donald Winnicott, Volume 4.* New York, NY: Oxford University Press.

Winnicott, Donald W., 1980 [1971]. *Playing and Reality.* Harmondsworth: Penguin.

_____, 2018 [1989]. *Psycho-Analytic Explorations.* London: Routledge.

Wittgenstein, Ludwig, 1958. *Philosophical Investigations.* Oxford: Blackwell.

Index

For Product Safety Concerns and Information please contact our EU
representative GPSR@taylorandfrancis.com
Taylor & Francis Verlag GmbH, Kaufingerstraße 24, 80331 München, Germany

www.ingramcontent.com/pod-product-compliance
Ingram Content Group UK Ltd.
Pitfield, Milton Keynes, MK11 3LW, UK
UKHW020143200926
475702UK00005B/881
* 9 7 8 1 0 3 2 3 3 0 0 1 3 *